PRAISE FOR

PANDEMIC

"Investigative science journalist [Sonia] Shah (*The Fever*, 2011) is at it again, and if the words, and beyond, in her latest book's subtitle don't grab a reader's attention, they should. This time, she is calling on global leaders, public and corporate, to pay attention to an impending public health emergency . . . Shah doesn't leave us wondering how this could happen in an age of proper sewage treatment and Purell. One-by-one she ticks off (no Lyme-disease pun intended) half a dozen conduits by which deadly microbes can spread faster than you can say methicillin-resistant *Staphylococcus aureus* (MRSA)—an infection, by the way, that affected her family. Yes, Shah is back and in rare form. And this time it's personal."   —Donna Chavez, *Booklist*

"Distinguished science journalist [Sonia Shah] takes us through the long, yet shrewdly articulated, journey that leads to human pandemics . . . Biting . . . Captivating . . . . This thought-provoking and well-documented book will contribute to spreading the word."   —Anne Cori, *Nature Microbiology*

"[An] absorbing, complex, and ominous look at the dangers posed by pathogens . . . Shah's warning is certainly troubling, and this important medical and social history is worthy of attention—and action."   —*Publishers Weekly*

"The ingredients for pandemics remain potent in a jet age with deforested lands, ever-growing cities, the consumption of bush meat and other exotic wild cuisine (from illegal 'wet markets'), antibiotic resistance, inadequate disease surveillance, and destructive cultural attitudes, ranging from abject fear to blame to indifference. Shah covers all of these aspects in vivid prose and through revealing eyewitness accounts. This is not fun reading, but it's necessary—one can only hope that it drives more effective surveillance and rapid response to tomorrow's plagues."   —*Kirkus Reviews*

Glenford Nuñez

A NOTE ABOUT THE AUTHOR

SONIA SHAH is a science journalist and the prizewinning author of *The Next Great Migration: The Beauty and Terror of Life on the Move*. Her writing on science, politics, and human rights has appeared in *The New York Times*, *The Wall Street Journal*, *Foreign Affairs*, and *Scientific American*, among other publications, and she has been featured on WNYC's *Radiolab*, NPR's *Fresh Air*, and TED.com, where her talk, "Three Reasons We Still Haven't Gotten Rid of Malaria," has been viewed by more than a million people around the world. Her book *The Fever* was long-listed for the Royal Society's Winton Prize for Science Books, and *Pandemic* was named a finalist for the Los Angeles Times Book Prize and a *New York Times Book Review* Editors' Choice.

PANDEMIC

TRACKING CONTAGIONS, FROM CHOLERA TO CORONAVIRUSES AND BEYOND

SONIA SHAH

PICADOR

FARRAR, STRAUS AND GIROUX

NEW YORK

Picador

120 Broadway, New York 10271

Printed in the United States of America
Published in 2016 by Sarah Crichton Books, an imprint of Farrar, Straus and Giroux
First Picador paperback edition, 2017
Second Picador paperback edition, 2020

The Library of Congress has cataloged the Sarah Crichton Books edition as follows:

Shah, Sonia.
   Pandemic : tracking contagions, from cholera to Ebola and beyond / Sonia Shah.
      p. cm.
   Includes index.
   ISBN 978-0-374-12288-1 (hardcover)
   ISBN 978-0-374-70874-0 (e-book)
   1. Communicable diseases—Epidemiology—History.   2. Public health surveillance. I.
Title.
   RA643 .S52 2016
   362.1—dc23   2015010246

Picador Paperback ISBN: 978-1-250-79324-9

Designed by Jonathan D. Lippincott

Our books may be purchased in bulk for promotional, educational, or business use. Please
contact your local bookseller or the Macmillan Corporate and Premium Sales Department at
1-800-221-7945, extension 5442, or by e-mail at MacmillanSpecialMarkets@macmillan.com.

Picador® is a U.S. registered trademark and is used by Macmillan Publishing
Group, LLC, under license from Pan Books Limited.

For book club information, please visit facebook.com/picadorbookclub or
e-mail marketing@picadorusa.com.

picadorusa.com • instagram.com/picador
twitter.com/picadorusa • facebook.com/picadorusa

1 3 5 7 9 10 8 6 4 2

pandemic: from the Greek *pan* "all" + *dēmos* "people."
A disease outbreak that spreads throughout an entire country, continent, or the whole world.

But what does it mean, the plague? It's life, that's all.

—Albert Camus

CONTENTS

PREFACE TO THE 2020 PAPERBACK EDITION

When the pandemic finally arrived, no one could quite believe it.

For weeks during the winter of 2020, experts at the World Health Organization insisted that the rapidly spreading contagion was no pandemic. Call it an "unprecedented outbreak," WHO experts said, as the virus slipped out of Wuhan, China. Or say it's a "public health emergency," they declared, as it reached Greece, five thousand miles west of Wuhan, and New Zealand, six thousand miles to the southeast.[1] By the time the planet's top public health officials admitted what the microbial tidal wave engulfing us really was—by then, commentators had taken to calling it "the p word"[2]—the virus had infiltrated the bodies of more than a hundred thousand people around the globe.[3] Millions more would suffer the same fate.

The earth was torn open to receive the bodies of the dead. On Hart Island, less than a mile offshore of the Bronx, aerial footage captured white-suited figures digging long, shallow trenches. They'd be filled with unclaimed corpses from the city's morgues—which had begun to overflow into refrigerated trucks—encased in pine coffins, the names of the dead scrawled on the lids.[4] Similar fissures appeared in satellite images taken eighty miles south of Tehran. Experts presumed the gashes in the earth were mass graves, situated next to piles of powdery lime required to mitigate the odor of decay. From space, they looked like wisps of cotton.[5]

Leaders around the world professed disbelief at the unfolding carnage. It's "unlike anything in our lifetime," headlines blared.[6] "Nobody has ever seen anything like this before," said U.S. President Donald Trump.[7] In *The New York Times*, one commentator noted that the virus's arrival was as unfathomable, and its presence as uncontrollable, as a natural disaster or an act of terrorism.[8] For many, even that assessment seemed too mundane. For them, the virus could only be understood as something altogether beyond nature. It must have originated in a lab, nearly a quarter of those who responded to a poll in the United States said.[9]

Shocked reporters unearthed old prognostications, showcasing them with awe and wonder. The National Security Council warned of a pandemic in a 2019 study, as *The Hill* reported in one headline. "Bill Gates predicted an epidemic long before the coronavirus," the *Los Angeles Times* announced, referring to his 2015 TED Talk on the subject.[10] It was "prophetic," *Rolling Stone* magazine added.[11] Predicting a pandemic, they implied, was akin to predicting a lightning strike, a fury that appears as if out of nowhere. How could they have known? Did they have a crystal ball that allowed them to see into the future? Some secret power that allowed them to predict the unpredictable?

With this unprecedented and unusual event holding the planet in its thrall, observers declared that a reckoning was on the horizon. The pathogen had exposed all our hidden soft spots and long-ignored problems: the lopsidedness that inflamed death rates among the poor and the marginalized; the laziness and corruption of local and national leaders who preferred to look the other way for too long; the fragility of our convoluted supply chains that were so easily disrupted, leaving some grocery aisles empty while fields were saturated with milk, dumped by dairy farmers whose thirsty customers were confined to quarantine. After the pandemic abated, there would be no hiding from the realities it had laid bare, commentators proclaimed. The pandemic's survivors would eventually emerge, chastened and forever altered. The pandemic would "change the world forever," *Foreign Policy* magazine asserted;[12] it would "change the world permanently," *Politico* wrote.[13] The virus would "change our lives forever," *The Washington Post* added.[14]

The contagion was a "portal," the novelist Arundhati Roy wrote, "a gateway between one world and the next."

⊙

For me, though, the most striking aspect of this pandemic is how eerily familiar it feels.

According to taxonomists, SARS-Cov2 hails not only from the same family of viruses as its fellow pandemic-causing pathogen SARS-Cov1 but also from the same species. The elder SARS virus emerged from the bodies of bats through the intermediary host of civet cats thanks to the peculiar opportunities furnished by the rapid expansion of the Chinese economy, which increased the probability of novel, close contact between people, bats, and other wild creatures in urban wet markets. The first SARS pandemic erupted in 2003. The lopsided balance it struck between contagiousness and deadliness caused it to burn out after killing close to eight hundred of the some eight thousand people it infected.

The virus-producing factory that led to the first SARS outbreak was never shut down, however. It was only a matter of time before another virus with a potentially more durable combination of attributes would emerge. Indeed, its younger sibling, just a little more contagious and a little less deadly, would best its ancestor by several orders of magnitude.

It erupted in a cluster of cases of severe pneumonia in Wuhan, China, in late December.[16] At first, local authorities refused to believe that these infections were anything out of the ordinary, censoring those who dared to suggest otherwise, just as government officials faced with novel outbreaks in the past had. By mid-January, when cases had already appeared in the United States, South Korea, Japan, and Thailand, the Wuhan city government held its annual Chinese New Year banquet. Forty thousand families gathered to share a celebratory meal—and the virions that clung to their hands and wafted outward with every breath. By the time government officials locked down the city a few days later, five million residents had already left, the coronavirus lodged in their throats, creeping steadily down into their lungs. They shared it with work colleagues and wedding guests and fellow cruise ship passengers, and the virus swept across the planet, the same way that nineteenth-century canal boats and steamships had washed cholera across the Atlantic and into North America, silently and efficiently, before anyone thought to notice. States slammed shut their borders and closed ports in its wake, with almost comical tardiness, just as officials in Wuhan had.

Belatedly, societies around the world woke up as if from a dream to the nightmare of the pandemic. Numbed by the scale of the threat and the mass mobilization that public health measures to contain it required, leaders around the world fell back upon worn metaphors. France was "at war" with the infection, its president said;[18] China would wage a "people's war," its president said;[19] Donald Trump would be a "war president."[20]

But a real war would require an invasive outsider vulnerable to acts of force—one that can be singled out, repelled, and destroyed. A war pits one opponent against another. The widely distributed virus, on the other hand, was already as well incorporated into bodies and societies as cotton thread woven into fabric. It would not "plot strategy," as one observer pointed out. It was "incapable of malice or fear."[21] But who or what would serve to stand in for the enemy was open to interpretation and outright manufacture. Some took a microscopic approach, targeting every last viral particle for chemical annihilation. They wiped down their cereal boxes with bleach and wore airtight scuba gear to the grocery store. In Jamaica, the enemy was a bus passenger who happened to sneeze—and who was seized, beaten, and thrown onto the road by his fellow passengers.[22] In Australia, it was a man who suffered cardiac arrest on the street outside a Chinese restaurant. Convinced they'd glimpsed the invisible microbial enemy inside him, the assembled crowd watched him die rather than provide lifesaving CPR.[23] Across Europe, mobs convinced that electromagnetic signals spread the virus vandalized and burned down cell towers.[24]

During the cholera pandemics of the nineteenth century, doctors were stoned in the streets and quarantine hospitals burned to the ground by angry crowds. During the COVID-19 pandemic, doctors in upstate New York were spat on and nurses emerged from late-night shifts to find the tires on their cars slashed. In Indore, India, a health-care worker in light blue scrubs walking down a narrow alley was attacked and chased by enraged locals.[25] In the White House, a frustrated and overwhelmed president blamed the World Health Organization.[26]

While fingers were pointed at the Chinese, at wild animals, at a lab somewhere, at political enemies, at shadowy international authorities, the virus quietly spread, undeterred.

☉

I write these words in the depths of quarantine, at a moment of maximum dread and despair. The toll of lives and livelihoods lost to the coronavirus mounts, inexorably. The much-anticipated changed world waiting for us on the other side of the pandemic remains blurry and dim; its contours just barely visible through a thick blanket of terror and hope. Whether it will arrive remains an open question.

In the past, pathogens have ripped through societies, exploiting our politics and ways of living, moving, and connecting with one another as efficiently and fatally as the novel coronavirus has today—some even more so. But even as they left deep imprints and jagged scars on our bodies and societies, we did not change our ways of life to shut them out, even when we could. On the contrary, almost as soon as the heat of contagion lifted, we resumed doing the same things we'd done before. Then, as now, we imagined pandemics to be as unfathomable and unpredictable **as being struck by lightning**. We cast them as acts of foreign aggression. We did not grapple with our complicity in their spread.

We did not make space for them in our historical memory, and by denuding pandemics of their social character, deprived them of history. We became their unwitting allies, and they came back, again and again. But it doesn't have to be so.

The pages that follow tell the story of pandemics through the lens of human agency. It's one in which the future of pandemics, like their past, lies entangled with our own. We hold it in our hands.

Baltimore, Maryland
April 17, 2020

PANDEMIC

INTRODUCTION: CHOLERA'S CHILD

Cholera kills people fast. There's no drawn-out sequence of progressive debility. The newly infected person feels fine at first. Then half a day passes, and cholera has drained his or her body of its fluids, leaving a withered blue corpse.

That's why, even after being infected, you could, say, eat a decent breakfast at your hotel, of sunny-side-up eggs and tepid juice. You could drive over dusty, potholed roads to the airport. You'd feel perfectly well enough to withstand the long queues there. Even as the killer silently brewed in your gut, you'd push your bags through security, perhaps even pick up a croissant at the coffee shop and enjoy a brief respite in a cool molded-plastic chair at the gate before a crackly PA announced the boarding of your flight.

It would only be after you'd shuffled down the plane's aisles and found your lightly tattered upholstered seat that the stranger within would make itself known, in a deadly, explosive onslaught of excretion, and your trip overseas would be suddenly and cruelly curtailed. Without the benefit of modern medicine rapidly administered, you'd be faced with a fifty–fifty chance of survival.

Such was the fate of a passenger ahead of me in line for Spirit Air Flight 952, from Port-au-Prince, Haiti, to Fort Lauderdale, Florida, in the summer of 2013. At the moment that cholera overcame the man, the rest of us were crammed inside a sweltering hall between the gate and the airplane, preparing to board. We waited there while the plane underwent an emergency disinfection. The airline didn't tell us what had caused the sudden hour-long delay. When an airline worker sprinted out of the plane through the hall to gather more supplies, impatient passengers bombarded him with a chorus of questions. He yelled out over his shoulder, by way of explanation, "A man shit himself." In Haiti, in the midst of a devastating cholera epidemic, there was little doubt as to what had happened.

If the stricken man had been infected an hour or two later, and sickened after we'd all taken our seats, with our arms jostling next to his on narrow strips of shared armrests, our knees grazing his, our hands brushing the overhead bins he'd touched, the pathogen might have been able to roost inside our bodies, too. I had spent my trip traipsing around to cholera treatment clinics and cholera-struck neighborhoods to see cholera firsthand. This formidable pathogen had nearly joined me on my flight home.

⊙

The disease-causing microbe, or pathogen, that will cause the world's next pandemic lurks among us today. We don't know its name or where it comes from. But for now call it "cholera's child," because what we do know is that it will likely follow the path that cholera blazed.

Cholera is one of only a handful of pathogens—including bubonic plague, influenza, smallpox, and HIV—that in modern times have been able to cause pandemics, contagions that spread widely among human populations. Among them, it stands alone. Unlike the plague, smallpox, and influenza, cholera's emergence and spread were well documented from the beginning. Two centuries after it first emerged, it remains exceptionally potent, with an undiminished power to cause death and disruption, as its display on Flight 952 showed. And unlike relative newcomers like HIV, cholera's an old hand at pandemics. So far, it's caused seven, the latest hitting Haiti in 2010.

Today cholera is known primarily as a disease that affects impoverished countries, but that wasn't always so. In the nineteenth century, cholera struck the most modern, prosperous cities in the world, killing rich and poor alike, from Paris and London to New York City and New Orleans. In 1836, it felled King Charles X in Italy; in 1849, President James Polk in New Orleans; in 1893, the composer Pyotr Ilyich Tchaikovsky in St. Petersburg. Over the course of the nineteenth century, cholera sickened hundreds of millions, killing more than half of its victims. It was one of the fastest-moving, most feared pathogens in the world.[1]

The microbe that causes the disease, *Vibrio cholerae*, was introduced into human populations during the era of British colonization of the South Asian hinterlands. But it was the rapid changes of the Industrial Revolution that provided the opportunities that turned the microbe into a pandemic-causing pathogen. New modes of travel—steamships, canals, and railways—ferried *Vibrio cholerae* deep into Europe and North America. The crowded and unsanitary conditions of rapidly growing cities allowed the bacteria to efficiently infect scores at a time.

Repeated epidemics of cholera posed a potent challenge to the political and social institutions of the societies it afflicted. Containing the disease called for a level of international cooperation, effective municipal governance, and social cohesion that had yet to be forged in newly industrialized cities and towns. Discovering its cure—clean water—required that doctors and scientists transcend long-held conventional wisdom about health and the spread of disease.

It took nearly a hundred years of deadly cholera pandemics for cities like New York, Paris, and London to rise to cholera's provocations. To do it, they had to remake the way they housed themselves, managed their drinking water and their waste, governed the public's health, conducted international relations, and understood the science of health and disease.

Such is the transformative power of pandemics.

⊙

The medical and public-health advances developed to contain nineteenth-century pathogens like cholera were so effective that for most of the twentieth century, the conventional wisdom among epidemiologists, medical historians, and other experts was that developed societies had vanquished infectious diseases for good. Western society had achieved "the virtual elimination of infectious diseases as a significant factor in social life," the virologist Sir Macfarlane Burnet wrote in 1951.[2] "To write about infectious disease," he added in 1962, "is almost to write of something that has passed into history."[3] The average American who survived for around five decades at the beginning of the twentieth century lived for nearly eight by century's end.[4]

According to the popular theory of "epidemiological transitions," first articulated by the Egyptian scholar Abdel Omran, the demise of infectious diseases in wealthy societies was an inevitable result of economic development. As societies prospered, their disease profile shifted. Instead of being plagued by contagion, they suffered primarily from slow-moving, chronic, noncommunicable conditions, like heart disease and cancer.

I confess to once being a true believer in this theory. I knew from visiting places like the south Mumbai ghetto where my father had grown up that societies that suffered significant burdens of infectious diseases were indeed crowded, unsanitary, and impoverished. We stayed in south Mumbai every summer, crammed with relatives into two-room flats in a dilapidated tenement building. Like the hundreds of other residents, we flung our waste into the courtyard, carried our own water in aging plastic buckets to shared latrines, and fitted two-foot boards over the thresholds to keep out the rats. There—as in other crowded, waste-ridden, poorly plumbed societies—infection was a constant reality.

But at the end of every summer, we boarded planes back home, and it seemed like we'd left behind that contagion-ridden way of life for good, tracing the path my parents had blazed when they'd first left India for New York, plastic-framed medical degrees in their bags. In the American towns where we lived, where drinking-water supplies had been cleaned up; sewage contained, treated, and distantly disposed of; and a public-health infrastructure built, infectious diseases were a problem that had been solved.

But then, thanks to the same conditions that brought cholera to the shores of New York City, Paris, and London in the nineteenth century, writ large, the microbes staged their comeback. Development in once remote habitats introduced new pathogens into human populations. A rapidly changing global economy resulted in faster modes of international travel, offering these pathogens new opportunities to spread. Urbanization and the growth of slums and factory farms sparked epidemics. Like cholera, which benefited from the Industrial Revolution, cholera's children started to benefit from its hangover: a changing climate, thanks to the excess carbon in the atmosphere unleashed by centuries of burning fossil fuels.

The first new infectious disease that struck the prosperous West and disrupted the notion of a "postinfection" era, the human immunodeficiency virus (HIV), appeared in the early 1980s. Although no one knew where it came from or how to treat it, many commentators exuded certainty that it was only a matter of time before medicine would vanquish the upstart virus. Drugs would cure it, vaccines would banish it. Public debate revolved around how to get the medical establishment to move quickly, not about the dire biological threat that HIV posed. In fact, early nomenclature seemed to negate the idea that HIV was an infectious disease at all. Some commentators, unwilling to accept the contagious nature of the virus (and willing to indulge in homophobic scapegoating) declared it a "gay cancer" instead.[5]

And then other infectious pathogens arrived, similarly impervious to the prevention strategies and containment measures we'd long taken for granted. Besides HIV, there was West Nile virus, SARS, Ebola, and new kinds of avian influenzas that could infect humans. Newly rejuvenated microbes learned to circumvent the medications we'd used to hold them in check: drug-resistant tuberculosis, resurgent malaria, and cholera itself. All told, between 1940 and 2004, more than three hundred infectious diseases either newly emerged or reemerged in places and in populations that had never seen them before.[6] The barrage was such that the Columbia University virologist Stephen Morse admits to having considered the possibility that these strange new creatures hailed from outer space: veritable Andromeda strains, raining down upon us from the heavens.[7]

By 2008, a leading medical journal acknowledged what had become obvious to many: the demise of infectious diseases in developed societies had been "greatly exaggerated."[8] Infectious pathogens had returned, and not only in the neglected, impoverished corners of the world but also in the most advanced cities and their prosperous suburbs. In 2008, disease experts marked the spot where each new pathogen emerged on a world map, using red points. Crimson splashed across a band from 30–60° north of the equator to 30–40° south. The entire heart of the global economy was swathed in red: northeastern United States, western Europe, Japan, and southeastern Australia. Economic development provided no panacea against contagion: Omran was wrong.[9]

As this realization ripples through the medical establishment, the power of microbes—that is, the army of minute organisms too small to see with the naked eye, such as bacteria, viruses, fungi, protozoa, and microscopic algae—looms large. Instead of conquest, today infectious-disease experts speak of their diminishing odds, of the possibility that perhaps even the cancers and the mental illnesses that we'd once chalked up to lifestyle and genetics were actually the work of other untamed microbes.[10] The old talk of mastery has vanished. "You hear this analogy that we have to win this war against microbes," said the UCLA infectious-disease expert Brad Spellberg to a room full of colleagues in 2012. "Really? They are so numerous that they collectively outweigh us by one-hundred-thousand-fold. I don't think so."[11]

As the number of new pathogens rises, so does the death toll. Between 1980 and 2000, the number of deaths pathogens caused in the United States alone rose nearly 60 percent. HIV was responsible for a large proportion of those deaths, but not all. Excluding HIV, the number of people felled by pathogens rose by 22 percent.[12]

Many experts believe that a cholera-like pandemic looms. In a survey by the epidemiologist Larry Brilliant, 90 percent of epidemiologists said that a pandemic that will sicken 1 billion, kill up to 165 million, and trigger a global recession that could cost up to $3 trillion would occur sometime in the next two generations.[13] So far, neither of the two pandemics caused by our newly emerged pathogens—HIV and H1N1—has been as fast and deadly as cholera. HIV is deadly, to be sure, but it spreads slowly; in 2009, H1N1 flu had spread quickly and widely but caused death in less than 0.005 percent of its victims.[14] But new pathogens have caused species-destroying pandemics among our fellow animals. Chytrid fungus, first spotted in 1998, now threatens the collapse of many species of amphibians. In 2004, pollinating insects began disappearing, victims of the still-mysterious colony collapse disorder. In 2006, white-nose syndrome, caused by the fungal pathogen *Pseudogymnoascus destructans*, started to decimate North American bats.[15]

Partly, this sense of an impending pandemic derives from the increasing number of candidate pathogens with the biological capacity to cause one. But it's also a reflection of the shortcomings in our public-health infrastructure, modes of international cooperation, and ability to maintain social cohesion in the face of contagion. The way modern societies have handled outbreaks of new diseases so far does not bode well. The Ebola virus broke out in a remote forest village in Guinea in early 2014. It would have been easy to contain using only the simplest, cheapest measures had it been squelched early on at its source. Instead, the virus, which had previously infected no more than a few hundred people at a time, in a single year spread into five neighboring countries, infected more than twenty-six thousand people, and would cost billions to contain.[16] Well-understood diseases that can be easily contained with drugs and vaccines escaped control even in the wealthy countries best situated to stanch them. An outbreak of vaccine-preventable measles that began in Disneyland over winter holidays in 2014 spread into seven states, exposing thousands to the contagion. Between 1996 and 2011, the United States experienced fifteen such outbreaks.[17]

Which, if any, of the new pathogens will cause the next pandemic in humans remains to be seen. By the time I boarded that plane in Haiti, I'd encountered a few of the contenders myself.

⊙

In 2010, my two sons, age ten and thirteen, were like walking scabs. With legs bare beneath flimsy athletic shorts, they kicked torn-up soccer balls on the tarmac, leapt off bridges into the rocky streambed behind the house, scuffled on the rough slate floor.

I hadn't registered the Band-Aid that my older son had affixed to his knee that spring. By the time he began complaining about it, the bandage's edges had started to fray and the exposed adhesive had captured a few days of grit. He said his knee hurt, but that was easy to explain away. The purported scab was located over his kneecap, after all, and it wasn't likely that he'd remained still long enough for it to ossify. A quick glance at the maroon blotch in the middle of the bandage provided sufficient evidence: this particular scab was being continually reopened. "Sure," I thought to myself, "that would hurt."

A few days later, he was wincing whenever he stood up. "Drama queen," I thought. The next morning, he came down to the kitchen limping.

We removed the Band-Aid. There was no scab. Instead, we found a mountain range of angry pus-filled boils. One peak summited at over an inch—an inch!—and had wept a sickly stream of liquid into the gummed-up bandage.

The pathogen that had created these abscesses, we soon learned, is called methicillin-resistant *Staphylococcus aureus*, aka MRSA ("mursa" as experts pronounce it). It's an antibiotic-resistant bacterium that first emerged in the 1960s and by 2010 was killing more Americans than AIDS.[18] The pediatrician, usually so jovial, turned sternly businesslike after one look at my son's knee. She whipped off a barrage of prescriptions even before the lab results came in: the heavy-duty antibiotic clindamycin and the old standby Bactrim, plus a brutal regime in which we'd have to force the pus out of the boils using hot compresses and viselike squeezing. This would be both excruciatingly painful, since the layer of pus extended deep under the tissue (tears ran down my son's face at the thought of it), and fraught, as scores of MRSA bacilli teemed inside the pus. Each drop would have to be meticulously captured and disposed of, lest it find its way into a microscopic fissure in our skin or, worse, embed itself in our rugs, sheets, couches, or counters, where it could lie in wait for up to a year.[19]

After weeks of squeezing and drugs, the infection seemed tamed. "He was lucky," a leading microbiologist said to me. "He could have lost his leg."[20] But summoned back to the pediatrician's office for a follow-up, we were told we hadn't seen the last of this unpredictable, hard-to-control new pathogen.

Whole families come down with MRSA, continually reinfecting each other for years, the pediatrician told us. I knew, having done some research by then, that this bug could kill, too. But none of the various physicians we visited knew how to prevent the infection from reoccurring or from spreading from my son to the rest of the family. One doctor recommended twice-weekly twenty-minute baths in bleach solution. "It's not a beauty treatment," he added, as if any clarification were needed. This, he said, we should continue until we were certain that there would be no more episodes—that is to say, for months, or even years. Another recommended the same treatment but with different specifics, namely, that the bath should have ½ cup of bleach in it. The doctor provided no details on duration or frequency, and in my shell-shocked state, I neglected to ask.

This lack of clear consensus, the open-ended time frame, and the repellent nature of the treatment began to shake our resolve. We started to wonder: Are they making it up? At the time, there'd been only one study on the effectiveness of bleach treatment, conducted in 2008. It showed that moderately concentrated bleach baths could "decolonize" material of MRSA. But how long-lasting the effect was, whether it would work on human skin as it did on the material used in the study, and, most important, whether it would make any difference in the frequency of MRSA infections one might get, nobody knew. Perhaps MRSA lived inside one's body, or victims were somehow primed to pick it up or become infected by it from other sources, in which case the bleaching would make no difference at all. And maybe, as my husband pointed out, the same result, such as it was, could be had by regular swimming in the MRSA-neutralizing, highly chlorinated waters of our neighborhood swimming pool. Or by regularly exposing our skin to sunlight.

Medicine's uncertainty about how to cope with this upstart offended my sensibilities. As the child of medical professionals (a psychiatrist and a pathologist), I'd grown up with the idea that medicine could solve all ills. How had the sureties of the past so quickly devolved into "perhaps" and "maybe"?

Adding to my sense of unease was the memory of an episode that had transpired the year before our initiation into life with MRSA. In 2009, a new kind of influenza virus, called H1N1, had arrived in the local elementary and middle schools. I had jostled at the clinic with scores of harried parents for a chance to get my kids vaccinated with the H1N1 shot. But H1N1 had come on too fast, too strong, and there wasn't enough vaccine. By the time my kids got the shot, it was too late; influenza (presumably H1N1, since it was the dominant strain in circulation that winter) had already started to incubate in their bodies. For days, two unstoppable boys lay utterly still, as their bodies burned with 103° fevers to repel the virus. As with MRSA, there was nothing to do, nothing to offer them. Finally, they recovered, although more than half a million others around the world died of H1N1—including more than twelve thousand in the United States. The rest of that season my sons' soccer car pools filled with the sound of a gaggle of boys emitting identical postflu hacking coughs.[21]

And then, within months of H1N1 and MRSA's incursions into my household, cholera washed over Haiti, where it hadn't been seen in more than a century.

⊙

This quick succession of events convinced me that the strange new infections we'd experienced were not isolated, circumstantial events but

part of a larger, global phenomenon. Having spent several years reporting on one of humankind's oldest pathogens, malaria, my interest was
immediately piqued. Most of the time, the story of pandemic disease begins when pathogens are already entrenched in populations, exacting
their pounds of flesh. The backstory of how they got there and where they
came from has to be pieced together from disparate clues and signs, an
especially challenging task when the subject is dynamic and constantly
evolving. And yet it's the backstory that is the most important one of all,
for it gives us the knowledge we need to prevent pandemics from taking
hold in the first place. The arrival of a spate of new pathogens provided
an opportunity to capture that backstory in real time. The obscure
mechanisms and pathways that turn microbes into pandemic-causing
pathogens could be tracked firsthand.

But I struggled with the question of how to do it. One approach
would have been to pick one emerging pathogen and track its development. For me, that seemed both risky and mercenary. Which one to
choose? While the overall risk of a pandemic may be rising, there's no
telling which of our emerging and reemerging pathogens, if any, will
cause one. I could make an educated guess—others have—but odds are
that guess would be wrong. Most emerging pathogens won't cause
pandemics. That's just a matter of math: very few pathogens do.

Another approach would have been to delve into the history of a
pathogen that has already mastered the business of causing pandemics.
That's a safer strategy but still would provide only a partial glimpse into
what's happening now. As fascinating as the stories of cholera or smallpox
or malaria are, each is necessarily rooted in its time and place. Plus, there's
an inherent paradox: the better, more detailed history one provides, the
more likely it is that the conditions that led to a historical pandemic will
come to seem unique and therefore tangential to the story of tomorrow's
pandemic.

I was idly browsing through papers about emerging diseases when I
stumbled upon a 1996 *Science* paper by the microbiologist Rita Colwell.
It was an adaptation of an address she'd given to the American Association for the Advancement of Science. In her talk, Colwell had posited
what she called the Cholera Paradigm: the idea that inside the story of
cholera, her longtime specialty, were all the clues required to understand the primary drivers behind other emerging diseases. It occurred to

me then that what I needed to do was to essentially combine the two approaches I'd previously dismissed in isolation. By telling the stories of new pathogens through the lens of a historical pandemic, I could show both how new pathogens emerge and spread, and how a pathogen that had used the same pathways had already caused a pandemic. The path from microbe to pandemic would be illuminated in the overlap, where two dim beams intersected.

And so I set off for the slums of Port-au-Prince, the wet markets of south China, and the surgical wards of New Delhi, in search of the birthplaces of pathogens old and new. I delved into the history of pandemics, in the written record as well as the one etched into our genomes. I tapped fields that ranged from evolutionary theory and epidemiology to cognitive science and political history, as well as my own idiosyncratic story.

What I found is that as similar as today's pace of economic, social, and political change is to that of the nineteenth century's era of industrialization, there's an important difference. In the past, the forces that drove pandemics were obscure to their victims. In the nineteenth century, people carried cholera across the seas on their ships and canals, allowed it to spread in their crowded slums and through their commercial transactions, and made its symptoms more deadly with their medicines without knowing how or why. Today, as we stand on the cusp of the next pandemic, the multistage journey from harmless microbe to pandemic-causing pathogen is no longer invisible. Each stage can be laid bare for all to see.

This book tracks that journey, from the wilds of colonial South Asia and the nineteenth-century slums of New York City to the jungles of Central Africa and the suburban backyards of the East Coast today. It begins, for cholera and its progeny, in the bodies of the wild animals around us.

ONE

THE JUMP

In search of the birthplace of new pathogens, I set out on a cool rainy day in early 2011 to find a wet market in Guangzhou, the capital of the southern Chinese province of Guangdong.

Wet markets are open-air street markets where vendors sell live animals captured from the wild to consumers to slaughter and consume. They service the Chinese taste for what's called *yewei*, or "wild," cuisine, in which exotic animals from snakes and turtles to bats are prepared into special dishes.[1]

It was in a wet market in Guangzhou that the virus that nearly caused a pandemic in 2003 was born. This particular virus normally lived inside horseshoe bats. It was a kind of coronavirus, a family of viruses that mostly cause mild respiratory illnesses. (In humans, they're responsible for about 15 percent of all cases of the common cold.) But the virus that was hatched at the wet market in Guangzhou was different.[2]

From the horseshoe bats, it had spread into other wild animals caged nearby, including raccoon dogs, ferret badgers, snakes, and palm civets. As the virus spread, it mutated. And in November 2003, a mutant form of the horseshoe bat virus started infecting people.

Like other coronaviruses, the virus colonized the cells lining the respiratory tract. But unlike its more mild brethren, the new virus tinkered with the human immune system, disrupting infected cells' ability to warn neighboring cells of the viral intruder in the body. As a result, in about a quarter of the infected, what started off seeming like flu rapidly escalated into life-threatening pneumonia as infected lungs filled with fluid and starved the body of oxygen. Over the following months, the virus sickened more than eight thousand with what came to be known as SARS, for severe acute respiratory syndrome. Seven hundred and seventy-four people perished.[3]

The SARS virus vanished after that. Like a brightly burning star, it used up all its available fuel, killing people too quickly to spread any farther. After scientific experts fingered wet markets as the hatcheries that birthed the strange new pathogen, Chinese authorities cracked down on the markets. Many closed. But then a few years passed and wet markets came back, albeit in smaller and more furtive form.

We'd been told there was a wet market somewhere around Zengcha Road, a traffic-clogged four-lane road that runs under a belching highway in Guangzhou. After walking around in circles for a bit, we stopped to ask a uniformed guard for directions. He laughed mirthlessly. The wet market was closed down six years ago, he said, after the SARS epidemic. But then, without pause, he grabbed the hem of a worker passing by and tugged on it, instructing us to reask our question, this time to the worker. We did, and the worker told a different story: go down around the other side of the building, he said, as the guard listened approvingly. We "may" find "some people" selling "some things."

As we turned the corner, the smell hit first, pungent, musky, and damp. The wet market consisted of a series of garage-like stalls lining a cement walkway. Some had been fashioned into office-cum-bedroom-cum-kitchens, in which the animal traders, bundled up for the weather, were passing the time waiting for customers. In one stall, three middle-aged men and a woman played cards on a folding table; in another, a bored-looking teenage girl watched a television bolted to the wall. As we walked in, a man flung the dregs of his soup bowl into the shallow gutter between the stalls and the walkway, a family of eight huddled around steaming bowls of hot pot behind him. A few minutes later, he reappeared to vigorously blow his nose into it.

Ignored entirely were the goods that we'd come to see: the caged wild animals that had been captured and acquired from other traders, in a long supply chain extending deep into China's interior and as far afield as Myanmar and Thailand. A thirty-pound turtle in a white plastic bucket sat desultorily in a puddle of gray water next to cages of wild ducks, ferrets, snakes, and feral cats. Row after row of animals who'd rarely if ever encounter each other in the wild were here, breathing, urinating, defecating, and eating next to each other.

The scene was remarkable in several ways that might explain why SARS had begun there. One was the unusual, ecologically unprecedented conglomeration of wild animals. In a natural setting, horseshoe bats, which live in caves, never rub shoulders with palm civets, a kind of cat that lives in trees. Neither would normally come within spitting distance of people, either. But all three came together in the wet market. The fact that the virus had spread from bats into civet cats had been especially critical to SARS's emergence. The civet cats were, for some reason, especially vulnerable to the virus. This gave the virus the opportunity to amplify its numbers, like a whistle in a tunnel. With increased replication came increased opportunities to mutate and evolve, to the extent that it evolved from a microbe that inhabited horseshoe bats to one that could infect humans. Without that amplification, it's hard to say whether the SARS virus would have ever emerged.

We approached one vendor in a stall lit by a single bare bulb. Behind him, on a sagging shelf, was a smudged, gallon-size glass jar packed with snakes floating in some kind of brine. As my translator Su engaged in small talk with the vendor, two women appeared and flung white cloth sacks on the floor at my feet. Inside one, a tangle of thin brown snakes slithered over each other. In the other, a single, much larger, violently jerking snake hissed. Clearly, this snake was perturbed. Through the sheer fabric I could see that the snake's head had a wide hood, which meant it was a cobra.

While those two facts sank in, the man and the two women, who had thus far not acknowledged my presence, turned to face me with some urgency in their expressions. Su translated their question: Exactly how many people do I plan to feed with this snake?

I stammered "ten" and turned away, flustered. A few minutes later, a woman approached us with another question. She gestured at me and, politely hiding a smirk behind her hand, asked Su whether it was true that foreigners like me ate turkeys. For her, I was the one with the strange eating habits.

◉

Cholera also started out in the bodies of animals. The creatures that harbored cholera live in the sea. They're a kind of tiny crustacean called copepods. They're about a millimeter long, with teardrop-shaped bodies and a single bright-red eye. Since they can't swim, they're considered a kind of zooplankton, drifting in the water and delaying gravity's pull to the depths with long antennae splayed outward like wings on a glider plane.[4] Though they're not talked about much, they're actually the most abundant multicellular creatures on Earth. A single sea cucumber might be covered with over two thousand copepods, a single hand-size starfish with hundreds. In some places, the copepods are so thick that the water turns opaque, and in a single season each one may produce nearly 4.5 billion offspring.[5]

Vibrio cholerae are their microbial partners. *V. cholerae* is a microscopic, comma-shaped species of bacteria from the genus *Vibrio*. Although *V. cholerae* can live on its own, free-floating in the water, it collects most lushly in and on copepods, sticking to copepods' egg sacs and lining the interior of their guts. Vibrio bacteria performed a valuable ecological function there. Like other crustaceans, copepods encase themselves in crusty exteriors made of a polymer called chitin (pronounced "kite-in"). Several times during their lifetimes, they shed their outgrown skins like snakes, discarding 100 billion tons of carapaces annually. Vibrio bacteria feed on this abundance of chitin, collectively recycling 90 percent of the ocean's excess chitin. Were it not for them, copepods' mountain of exoskeletons would starve the ocean of carbon and nitrogen.[6]

Vibrio bacteria and copepods proliferated in warm, brackish coastal waters, where fresh and salty waters met, such as in the Sundarbans, an expansive wetlands at the mouth of the world's largest bay, the Bay of Bengal. This was a netherworld of land and sea long hostile to human penetration. Every day, the Bay of Bengal's salty tides rushed over the Sundarbans' low-lying mangrove forests and mudflats, pushing seawater as far as five hundred miles inland, creating temporary islands of high ground, called chars, that daily rose and vanished with the tides. Cyclones, poisonous snakes, crocodiles, Javan rhinoceros, wild buffalo, and even Bengal tigers stalked the swamps.[7] The Mughal emperors who ruled the Indian subcontinent up until the seventeenth century prudently left the Sundarbans alone. Nineteenth-century commentators called it "a

sort of drowned land, covered with jungle, smitten by malaria, and infested by wild beasts," and possessed of an "evil fertility."[8]

But then, in the 1760s, the East India Company took over Bengal
and with it the Sundarbans. English settlers, tiger hunters, and colonists
streamed into the wetlands. They recruited thousands of locals to chop
down the mangroves, build embankments, and plant rice. Within fifty
years, nearly eight hundred square miles of Sundarbans forests had been
razed. Over the course of the 1800s, human habitations would sprawl
over 90 percent of the once untouched, impenetrable, and copepod-rich
Sundarbans.[9]

Contact between human and vibrio-infested copepod had probably
never been quite so intense as in these newly conquered tropical wetlands. Sundarbans farmers and fishermen lived in a world semisubmerged
in the half-salty water in which vibrio bacteria thrived. It wouldn't have
been particularly difficult for the vibrio to penetrate the human body. A
fisherman who splashed his face with water by the side of a boat, say, or
a villager drinking from a well corroded with a few ounces of floodwaters, could easily ingest a few invisible copepods. Each one might be
infested with as many as seven thousand vibrios.[10]

This intimate contact allowed *Vibrio cholerae* to "spill over" or "jump"
into our bodies. The bacteria wouldn't have found a particularly welcoming reception there, at first. Human defenses are designed to repel
such intrusions, from the acid environs of our stomachs, which neutralize most bacteria, and the competitive wrath of the microbes that
inhabit our gut to the constantly patrolling cells of the immune system.
But in time, *V. cholerae* adapted to the human bodies to which it was
repeatedly exposed. It acquired, for example, a long, hairlike filament
at its tail that improved its ability to bond to other vibrio cells. Endowed
with the filament, the vibrio could form tough microcolonies that could
stick to the lining of the human gut like scum on a shower curtain.[11]

Vibrio cholerae became what's known as a zoonosis, from the Greek
zoon for "animal" and *nosos* for "disease." It was an animal microbe that
could infect humans. But *V. cholerae* wasn't a pandemic killer yet.

⊙

As a zoonosis, *Vibrio cholerae* could infect only people who exposed
themselves to their "reservoir" animals, the copepods. It was a pathogen

on a leash, unable to infect anyone outside its limited purview. It had
no way of infecting anyone who wasn't exposed to copepod-rich waters.
While it could cause outbreaks, when multiple people exposed themselves to copepods simultaneously, for example, those outbreaks would
always be self-limited. They'd collapse on their own.

For a pathogen to cause a wave of sequential infections—an epidemic
or a pandemic, depending on how far the wave traveled—it must be
able to spread directly from one human to another. That is to say, its
"basic reproductive number" has to be greater than 1. The basic reproductive number (also known as R_0, or "R-naught," as the Anglophiles
pronounce it) describes the average number of susceptible people who
are infected by a single infected person (in the absence of outside interventions). Say you have a cold, and you infect your son and his friend
with it. If this hypothetical scenario were typical of the entire population, your cold's basic reproductive number would be 2. If you infect
your daughter as well, your cold's basic reproductive number would be 3.

This calculation is a critical one to make in an outbreak, for it immediately predicts its future course. If, on average, each infection results in
less than one additional infection—you infect your son and his friend,
but each of them infects no one else—then the outbreak will die out
on its own, like a population in which each family produces fewer
than two children. It doesn't matter how deadly the infection is. But if,
on average, each infection results in one more infection, then the outbreaks can theoretically continue indefinitely. If each infection results
in more than one additional infection, then the afflicted population is
facing an existential threat that requires immediate and urgent attention. It means that, in the absence of interventions, the outbreak will expand exponentially.

The basic reproductive number is, in other words, a mathematical
expression of the difference between a zoonotic pathogen and one that
has crossed the threshold to become a human one. The basic reproductive number of a zoonotic pathogen, which cannot spread from one infected person into another, is always less than 1. But as it refines its
attack on humans, its ability to spread among them improves. Once it
tips over the number 1, the pathogen has crossed the threshold and broken free of its reservoir animals. It's a bona fide human pathogen that is
self-sustaining in humans.

There are many mechanisms by which zoonotic pathogens can acquire the ability to spread directly between people, severing the cord that binds them to their reservoir animals. *Vibrio cholerae* did it by acquiring the ability to produce a toxin.

The toxin was the vibrio's pièce de résistance. Normally, the human digestive system sends food, gastric and pancreatic juice, bile, and various intestinal secretions to the intestines, where cells lining the gut extract nutrients and fluid, leaving behind a solid mass of excreta to expel. The vibrio's toxin altered the biochemistry of the human intestines such that the organ's normal function reversed. Instead of extracting fluids to nourish the body's tissues, the vibrio-colonized gut sucked water and electrolytes out of the body's tissues and flushed them away with the waste.[12]

The toxin allowed the vibrio to accomplish two things essential to its success as a human pathogen. First, it helped the vibrio get rid of its competitors: the massive torrent of fluids sloughed off all the other bacteria in the gut, so that the vibrio (clinging to the gut in its tough microcolonies) could colonize the organ undisturbed. Second, it assured the vibrio's passage from one victim to another. Even tiny drops of that excreta, on unwashed hands or contaminated food or water, could carry the vibrio to new victims. Now, so long as the vibrio could get into a single person and cause disease, it could spread to others, whether or not they exposed themselves to copepods or ingested the vibrio-rich waters of the Sundarbans.

The first pandemic caused by the new pathogen began in the Sundarbans town of Jessore in August 1817 after a heavy rainfall. Brackish water from the sea flooded the area, allowing salty copepod-rich waters to seep into people's farms and homes and wells. *V. cholerae* slipped into the locals' bodies and colonized their guts. Thanks to the toxin, *Vibrio cholerae*'s basic reproductive number, according to modern mathematical models, ranged from 2 to 6. A single infected person could infect as many as half a dozen others.[13] Within hours, cholera's first victims were being drained alive, each expelling more than fifteen quarts of milky-white liquid stool a day, filling the Sundarbans' streams and waste pits with excreta. It leaked into farmers' wells. Droplets clung to people's hands and clothes. And in each drop, vibrio bacteria swarmed, ready to infect a new host.[14]

The Bengalis called the new disease *ola*, for "the purge." It killed people faster than any other disease known to humankind. Ten thousand perished. Within a matter of months, the new plague held nearly two hundred thousand square miles of Bengal in its grip.[15]

Cholera had made its debut.

⊙

Given the ubiquity of microbes, it would seem that new pathogens could come from anywhere at all, surfacing from their hidden corners to encroach upon humankind from all directions. They could be living inside of us already, becoming pathogenic by exploiting newfound opportunities within, or could emerge from the inanimate environment, like the soil or the pores of rocks and ice cores, or from any number of other microbial niches.

And yet that's not how most new pathogens are born, because their entry into our bodies is not random. Microbes become pathogens by following the avenues we pave for them, and these avenues follow particular routes. Even though there are countless repositories of microbes that could become human pathogens, most new human pathogens are like *Vibrio cholerae* and the SARS virus: they originate in the bodies of other animals. More than 60 percent of our newly emerged pathogens originate in the furred and winged creatures around us. Some of these new pathogens come from domesticated animals, such as pets and livestock. Most—over 70 percent—come from wild animals.[16]

Microbes have been crossing between species and turning into new pathogens for as long as humans have lived among other species. Hunting animals and eating them, which exposes people to the internal tissues and fluids of other animals, provides a good opportunity. Being bitten by insects like mosquitoes and ticks, which ferry other animals' fluids into our bodies, can do the trick, too. These are ancient forms of intimate contact between *Homo sapiens* and other animals, dating back to our earliest years, and they brought us some of our oldest pathogens, such as malaria, ferried from the bodies of our fellow primates into ours by blood-sucking mosquitoes.

Because intimate contact between species must be prolonged in order for an animal microbe to turn into a human pathogen, historically we are victim to some animals' microbes more than others. Many more pathogens come from the bodies of Old World creatures, whom we've lived with for millions of years, compared to New World creatures, with whom we've been acquainted for just tens of thousands of years. A disproportionate number of human pathogens hail from other primates, who've bestowed upon us 20 percent of our most burdensome pathogens (including HIV and malaria), despite comprising just 0.5 percent of all vertebrates. It's also why so many human pathogens date back to the dawn of agriculture ten thousand years ago, when people started domesticating other species and living in prolonged, intimate contact with them. From cows, we got measles and tuberculosis; from pigs, pertussis; from ducks, influenza.[17]

But while animal microbes have been spilling over into humans (and vice versa) for millennia, it's historically been a rather slow process.

Not anymore.

⊙

Peter Daszak is the scientist who discovered that horseshoe bats were the reservoir of the SARS virus. He directs an interdisciplinary organization that investigates emerging diseases in people and wildlife. I met him one afternoon in his office in New York City. He fell into the business of disease hunting by happenstance, he says. As a kid growing up in Manchester, England, he'd wanted to become a zoologist. "My big thing is lizards," he says, gesturing toward his pet captive-bred Madagascan day gecko, motionlessly poised in a lighted glass tank next to the front door. At his university, though, all the research projects on the behavior of lizards were filled. Daszak had to settle for one on the diseases of lizards instead. "My God, that's boring," he thought at the time.[18]

That inquiry, however, turned him into one of the world's foremost disease hunters. Daszak was working at the Centers for Disease Control in the late 1990s when herpetologists started noticing sudden declines in amphibian populations around the world. Few experts suspected disease. Biologists at the time believed that pathogenic microbes never threatened the survival of their host populations. Such virulence was considered self-defeating: if a pathogen killed too fast or too many of its victims, there'd be none left for it to subsist upon. And so "they were coming up with all these standard theories" to explain the mass mortality in amphibians, Daszak remembers. They thought the culprit might be a pollutant or a sudden change in the climate. But Daszak suspected that a never-before-seen contagion was killing the amphibians. He had already discovered a disease that had led to the extinction of an entire species of tree snails from the South Pacific.

In 1998, he published a paper reporting that a fungal pathogen—what turned out to be *Batrachochytrium dendrobatidis*, or amphibian chytrid fungus for short—had caused the amphibian declines around the world. Most likely, the pathogen had spread thanks to the quickening pace of disruptive human activities, in particular the stepped-up demand for amphibians as pets and in scientific research.[19]

And he was struck by something else. Humans, too, were vulnerable to pathogens unleashed by the same accelerated, disruptive forces that brought chytrid fungus to the world's amphibians. As wetlands were paved over and forests were felled, different species came into novel, prolonged contact with each other, allowing animal microbes to spill over into human bodies. And these developments were proceeding at an unprecedented scale and speed around the world.

The road from animal microbe to human pathogen was turning into a highway.[20]

⊙

Take the southwest corner of the West African nation of Guinea. One of the world's most biodiverse forests once covered the region. Large tracts of undeveloped forests, being difficult for people to penetrate, had limited contact between forest animals and humans. Wild animals could live in the forest without encountering humans or human settlements.

That changed during the 1990s, as the Guinean forest was steadily destroyed. A wave of refugees descended upon the forest to escape a long, bloody, complex conflict between armies and rebel groups from neighboring Sierra Leone and Liberia. (At first they'd tried to settle in refugee camps in the forest area's central town of Guéckédou, but rebel groups and government soldiers repeatedly attacked their camps.)[21]

The refugees cut down trees to plant crops, build huts, and turn into charcoal. Rebel groups started logging the forest, too, selling timber to finance their battles.[22] By the end of the 1990s, the transformation of the forest could be seen from space. In satellite images from the mid-1970s, the jungles in Guinea bordering Liberia and Sierra Leone looked like a sea of green splattered with tiny islands of brown, where small clearings had been made for villages. Satellite images from 1999 showed a complete reversal: a sea of treeless brown, with tiny islands of green forest speckled in between. Of the entire region's original forests, only 15 percent remained.[23]

Just how this wide-scale deforestation affected the forest ecosystem has yet to be fully described. Many species that lived in the forest probably just disappeared when humans moved into their habitats. What is known is that some species stayed. They squeezed in, resorting to smaller patches of remaining stands of trees, in increasing proximity to human habitations.

Bats were among them. It stands to reason: bats are widely distributed and resilient creatures. Of 4,600 species of mammals on earth, 20 percent are bats. And, as a study in Paraguay found, certain bat species thrive in disturbed forests in even higher abundances than in intact ones.[24] Unfortunately, bats are also good incubators for microbes that can infect humans. They live in giant colonies of millions of individuals. Some species, like the little brown bat, can survive for as long as thirty-five years. And they have unusual immune systems. For example, because their bones are hollow, like those of birds, they don't produce immune cells in their bone marrow like the rest of us mammals do. As a result, bats host a wide range of unique microbes that are exotic to other mammals. And they travel around with these microbes over significant distances, because bats can fly. Some even migrate, traveling thousands of miles at a time.[25]

As the Guinean forest was chopped down, new kinds of collisions between bats and people likely occurred. Bats were hunted for meat, exposing hunters to microbe-laden bat tissue when the animals were slaughtered. Bats fed on fruit trees near human settlements, exposing local people to their saliva and excreta. (Fruit bats are notoriously messy eaters; their modus operandi is to pick off ripe fruit and suck out the juice, littering the ground below with saliva-covered, half-eaten fruits.)

At some point—nobody knows just when—a microbe of bats, the filovirus Ebola, started to spill over and infect people. In humans, Ebola causes hemorrhagic fever and can kill 90 percent of those it infects.[26] A study of blood samples collected from people in eastern Sierra Leone, Liberia, and Guinea between 2006 and 2008 revealed that nearly

9 percent had been exposed to Ebola: their immune systems had created specific proteins called antibodies in response to the virus.[27] A 2010 study of over four thousand people in rural Gabon, where there'd been no outbreaks of Ebola, similarly found that nearly 20 percent had been exposed to the virus.[28]

But nobody noticed. The ongoing conflict had severed supply routes and communication networks, leaving the refugees hiding in the jungle bereft of outside help. Even the most stalwart aid organizations such as Médecins Sans Frontières had been forced to withdraw. The isolation coupled with the violence compelled the United Nations to call the West African refugees' plight "the worst humanitarian crisis in the world."[29]

It wasn't until the political violence eased, in 2003, and the people hiding in the Guinean forest slowly reconnected with the rest of the world that the virus's presence became apparent. On December 6, 2013, Ebola virus sickened and killed a two-year-old child in a small forest village outside Guéckédou. Perhaps the toddler had played with a piece of fruit covered with bat saliva, fallen from a nearby tree. Perhaps the parents had been handling a recently slaughtered bat before picking up the child. It was probably not the first time someone in the Guéckédou area had encountered Ebola virus from a local bat. But this time, the people of Guéckédou were no longer as isolated as they'd been in the past. The virus was able to spread.

By February 2014, a health-care worker had ferried the virus to three other local forest villages. Within a month, at least four clusters of cases had been ignited in Guinea's forest region, triggering independent chains of transmission.[30]

By the time hospital officials and aid workers alerted the Ministry of Health and the World Health Organization to the outbreak in the Guinea forest in March 2014, the virus had already spread into Sierra Leone and Liberia.[31] Six months later, the virus had emerged in urban centers throughout the region, and the size of the epidemic was doubling every two to three weeks. According to calculations by modelers, each infected person infected at least one or two others, for a basic reproductive number that ranged from 1.5 to 2.5. In the absence of containment measures, the Ebola outbreak would grow exponentially.[32]

Ebola had caused outbreaks on the continent before. Sporadic, contained eruptions in remote villages in Central Africa had occurred since the 1970s, mostly during the transition between rainy and dry seasons, possibly connected to the fruiting of trees, and in the wake of the arrival of large numbers of migratory bats. But never before had the virus caused the devastation that it did in West Africa. The thousands of people infected with Ebola quickly overwhelmed the fragile economies and health-care infrastructures of the three most affected countries. "None of us experienced in containing outbreaks has ever seen, in our lifetimes," the World Health Organization's director-general Margaret Chan said, "an emergency on this scale."[33]

In September 2014, the Centers for Disease Control estimated that Ebola might sicken more than a million people across West Africa.[34] That estimate proved to be overblown, but many believed it possible at the time. Ebola had already wreaked catastrophic damage on our fellow primates, the gorillas and chimpanzees who feed on the same fruit trees as fruit bats. Over the course of the 1990s and early 2000s, Ebola had killed one-third of the world's gorilla population and nearly the same proportion of the world's chimpanzees. By the time the epidemic in Guinea, Sierra Leone, and Liberia finally started to ebb in early 2015, more than ten thousand people had perished.[35]

⊙

Ebola is the most dramatic of the new animal microbes spilling over into people from forest animals in Africa, but it is not the only one.

Monkeypox is a virus that lives in Central African rodents. It comes from the same genus of viruses as the now extinct variola, the virus that caused smallpox, which killed between 300 and 500 million people over the course of the twentieth century. In humans, monkeypox causes a disease clinically indistinguishable from smallpox, with characteristic raised lesions, or pox, across the body, particularly on the face and hands. Unlike smallpox, monkeypox is a zoonosis. But according to studies conducted by the University of California epidemiologist Anne Rimoin, it has started to spill over into humans with increasing frequency.[36]

Between 2005 and 2007, Rimoin tracked down cases of monkeypox that had occurred in fifteen remote villages in the Democratic Republic of Congo. She took blood samples from those who'd been infected and confirmed that monkeypox was indeed the culprit. When she tallied her numbers, she found that monkeypox infection in humans had grown twentyfold compared to the period between 1981 and 1986.[37]

A variety of factors help explain the increase. For one thing, intimate contact between rodents and humans has become more common. Thanks to forest destruction, more people live in and around the monkeypox-infected rodents of Central African forests.[38] Because wild game populations and local fisheries have collapsed, many hunt for bushmeat, including the rodents they might have once rebuffed. The cessation of vaccination against smallpox plays an important role, too. A global mass vaccination campaign that stamped out smallpox in the late 1970s had conferred lifelong immunity against smallpox's entire genus of viruses to its recipients, which included monkeypox. But that campaign had ended in DRC in 1980. Everyone born after that time is as vulnerable to monkeypox as nonvaccinated people were to smallpox centuries ago.[39]

For now, monkeypox is still tethered to the rodents—probably rope squirrels, Rimoin says, but it's unclear—that carry it. It's only occasionally able to spread directly from human to human. According to Rimoin's colleague, the ecologist James Lloyd-Smith, monkeypox's basic reproductive number in humans is somewhere between 0.57 and 0.96, just shy of the 1 required to graduate from a zoonotic to a human pathogen. The Central African populations it infects, after all, are relatively remote and spread out. There aren't that many people around for monkeypox to jump into.[40]

Fortunately, even if monkeypox does complete the journey from animal microbe to human pathogen, it's unlikely that it would have the impact that smallpox did. The vaccines and drugs developed to fight smallpox would likely also help contain human-adapted monkeypox outbreaks. But monkeypox, Rimoin says, is the devil we know. Being similar to smallpox, and manifesting itself in a distinctive disease that's hard not to notice, monkeypox is a spillover microbe that's relatively easy to track. Microbes that cause less immediately noticeable symptoms could easily travel the same spillover pathways, undetected. Some probably already have.

⊙

The emergence of the SARS virus is similarly the result of an abrupt expansion, in this case in the size of wet markets and the diversity of the strange panoply of animals they sell.

The SARS virus was not new. Nor were the practices that brought bats into proximity with people in southern China. The SARS virus "was probably there in bats for centuries," says the University of Hong Kong virologist Malik Peiris, whose team first isolated the virus.[41] And the *yewei* cuisine and wet markets that brought bats together with people in southern China were long-standing, too.

Yewei cuisine is part of a range of traditional cultural practices in China that draw wild animals closer so that people can tap into the animals' power, strength, and longevity. People keep wild animals as pets (or, for the aspirational, dye their domesticated dogs' fur to look like tigers and pandas) and mimic their postures in practices such as kung fu. Traditional medicine practitioners administer their body parts as remedies: tiger whiskers for toothaches, bear bile for liver disease, bat skeletons for kidney stones.[42] For people who consider wild animals precious natural resources—the rarer, wilder, and more exotic the more precious—consuming them is *bu*, restorative and stimulating for the body, endowing the consumer with a whiff of the animal's natural energy.[43]

But for many years, economic and geographic barriers limited the consumption of *yewei* cuisine in China, and with it the size of wet markets. China had troubled political relations with neighboring countries such as Thailand and Laos and Vietnam, where many of the most desirable exotic animals roamed, so their supply for consumption was thin and prices high. While the elites could afford to dine on braised bear paw with carp tongue, gorilla lips and pig brain in wine sauce, and leopard placenta steamed with camel hump and garnished with pear, ordinary folk made do with more ordinary fare, or hunted for their own wild game.[44]

Then, in the early 1990s, the Chinese economy started growing by 10 percent or more every year. Suddenly, a new class of young, aspiring, prosperous Chinese in booming cities had more money than they knew what to do with. Along with stocking up on Western luxury goods—Louis Vuitton sold more bags in China than anywhere else in 2011—they started demanding more *yewei* cuisine. New restaurants serving peacock, swan geese, and sea cucumber, along with other exotic creatures, sprang up across the region.[45] China reestablished trade with many of its Southeast Asian neighbors, allowing poachers and traders to plunge ever deeper into the countryside to meet the rising demand. They crammed their stocks of wild animals into ever larger wet markets, stacking cages of live animals from increasingly disparate locales across Asia next to each other, awaiting sale to *yewei*-hungry shoppers.[46]

It was only then, after the size and scale of wet markets grew, that a serendipitous sequence of events that could turn a virus of horseshoe bats into a human pathogen became probable.

The growing size of pig farms in Malaysia had similarly allowed a bat virus called Nipah to spill over into people. As pig farms in Malaysia have grown, they've increasingly abutted the forests where bats roost. This allows bats and pigs to come into novel, intimate contact. The pigs' troughs are situated near the fruit trees where bats roost. When bat excreta drop into those troughs, the pigs are exposed to bat microbes. At one particularly large pig farm, Nipah virus sickened so many pigs that it was able to spill over into the local farmers, killing 40 percent of those infected. Nipah virus also struck in South Asia and now erupts in Bangladesh nearly every year, killing 70 percent of the afflicted.[47]

These spillovers are happening not only in distant societies and in impoverished, tropical locales. They're also happening in cities at the center of the global economy such as New York and in the prosperous suburbs of the northeastern United States.

⊙

West Nile virus is a flavivirus of migratory birds named after the district in Uganda where it was first isolated in 1937. Migratory birds have probably been introducing the virus into the United States for decades, especially into New York City, which lies on the Atlantic flyway, one of the four primary migratory routes in North America. The virus can spill over from birds' bodies into ours when mosquitoes bite infected birds and then bite people.

But despite the repeated introduction of the virus and the enduring phenomenon of mosquito biting, West Nile virus did not break out in the United States until 1999, more than fifty years after it was identified.

That's because the diversity of the local bird population limited our exposure to it. Different bird species vary in their vulnerability to the virus. Robins and crows are especially susceptible. Woodpeckers and rails are not. Their feathery bodies act like a barrier. So long as the local bird population was diverse, including plenty of virus-repelling woodpeckers and rails, there wasn't much virus around. The chances that the virus would spill over from birds into humans were slim.

But avian biodiversity, like the biodiversity of other species, has plummeted, in the United States as elsewhere. Urban sprawl, industrial agriculture, and climate change, among other disruptions caused by human activity, steadily destroy bird habitats, reducing the number of species among us. But habitat destruction doesn't affect all species equally. Some species—the so-called specialist species—get hit especially hard. They're the ones, like monarch butterflies, salamanders, and woodpeckers and rails, that rely on exacting conditions and can't easily survive when those conditions change. When trees are felled and nesting grounds paved over, they're the ones that tend to disappear first. That means there's more food and territory around for the "generalist" species like robins and crows—the opportunistic, sharp-elbowed types that can live anywhere and eat anything. Their numbers skyrocket in the vacuum.

As avian diversity declined in the United States, specialist species like woodpeckers and rails disappeared, while generalist species like American robins and crows boomed. (Populations of American robins have grown by 50 to 100 percent over the past twenty-five years.)[48] This reordering of the composition of the local bird population steadily increased the chances that the virus would reach a high enough concentration to spill over into humans. At some point, a threshold was crossed. In the summer of 1999, West Nile infected over 2 percent of the population of Queens, more than eight thousand people, in New York City.[49] Once it took hold, it spread inexorably. Within five years, the virus had emerged in all forty-eight contiguous states. By 2010, an estimated 1.8 million people in North America had been infected from New York to Texas and California. Experts agree that West Nile is here to stay.[50]

The loss of species diversity in northeastern forests of the United States similarly allowed tickborne pathogens to spill over into humans. In the original, intact northeastern forests, a diversity of woodland animals such as chipmunks, weasels, and opossums abounded. These creatures imposed a limit on the local tick population, for a single opossum, through grooming, destroyed nearly six thousand ticks a week. But as the suburbs grew in the Northeast, the forest was fragmented into little wooded plots crisscrossed by roads and highways. Specialist species like opossums, chipmunks, and weasels vanished. Meanwhile, generalist species like deer and white-footed mice took over. But deer and white-footed mice, unlike opossums and chipmunks, don't control local tick populations. When the opossums and the chipmunks disappeared, tick populations exploded.[51]

As a result, tickborne microbes increasingly spill over into humans. The tickborne bacteria *Borrelia burgdorferi* first emerged in humans in an outbreak in Old Lyme, Connecticut, in the late 1970s. If left untreated, the disease it caused—Lyme disease—can lead to paralysis and arthritis among other woes. Between 1975 and 1995, cases increased twenty-five-fold. Today, three hundred thousand Americans are diagnosed with Lyme every year, according to estimates from the Centers for Disease Control. Other tickborne microbes are spilling over as well. Between 2001 and 2008, cases of tickborne *Babesia microti*, which causes a malaria-like illness, increased twentyfold.[52]

Neither West Nile virus nor *Borrelia burgdorferi* and its kin can spread directly from one person to another, yet. But they continue to change and adapt. And elsewhere, the reordering of wildlife species that precipitated their spillover into humans proceeds. Globally, 12 percent of bird species, 23 percent of mammals, and 32 percent of amphibians are at risk of extinction. Since 1970, global populations of these creatures have declined by nearly 30 percent. Just how these losses will shift the distribution of microbes between and across species, pushing some over the threshold, remains to be seen.[53]

⊙

Strains of my family's emerging pathogen, MRSA, come from animals. Pigs harbor MRSA. They pass it on to their handlers, and the bacteria appear in their slaughtered flesh sold at the supermarket, although whether people get infected from eating it is still an open question. A University of Iowa study found that 3 percent of meat samples collected from Iowa grocery stores carried MRSA. In the Netherlands, the strain of MRSA commonly found among pigs causes 20 percent of MRSA infections in humans.[54]

I've never lived near a pig farm. But I have been known to eat the flesh of pigs. It's not something I'm particularly proud of. I grew up in a strictly vegetarian household. My parents were both raised as Jains, a religion that preaches an extreme form of nonviolence. Its cardinal rule is not to harm any other living thing, not even by stepping on grass (you might smash an insect) or breathing in germs, which is why while praying, my Jain grandmothers tied white cotton masks over their mouths. My Jain aunt wouldn't even accept my youthful offer of a goldfish cracker, because of the sinful implication of its fishy shape. No, Jains are supposed to be kind and gentle to our fellow animals, by doing things like spooning sugar on anthills, or, as my grandfather used to do, visiting Jain-run animal sanctuaries to hand-feed the cows and sheep rescued from the slaughterhouse. Shamefully, my only nod to these admirable traditions is a reluctance to visit zoos, or to kill the confused flies and spiders and ants that end up in my kitchen.

A true Jain, of course, would never be complicit in the usurpation of wild habitat or the business of herding animals onto giant farms and markets, which drive animal microbes into human populations. That's not me. So perhaps there was a certain logic in what happened the year after MRSA infected my son: a visceral display of an emerging pathogen's ability to spread from person to person, the first, necessary prerequisite to becoming a pandemic pathogen.

My son's second bout of MRSA arrived a few months after the first, requiring another round of semitoxic antibiotics. He spiked a fever while on the meds and had to be sent home from school. I was out of town with the family car, so he had to walk home, while I spent the rest of the day rushing back, fretting about the fever. Was he reacting to the antibiotics, or had his MRSA grown impervious to them? How could we tell the difference? If he was reacting to the antibiotics, would there be any other effective medications he could use? One whole class of the drugs had already been ruled out when they had caused an angry rash. If, on the other hand, MRSA had broken through the antibiotics, did that mean it would now lodge deeper into his body, into his tissues and organs? I'd read about the cases of healthy young people, like a twenty-one-year-old college student in Minnesota who'd developed a MRSA infection in the lungs, and a seven-year-old whose MRSA infection in her right hip had spread to her lungs. They both died.[55]

A third MRSA infection appeared on the inside of his elbow after another few months. By this time, there was no doubt: MRSA lived inside his body. There was no fissure in this protected bit of skin that would have allowed an external invader to creep in. The infection appeared from within. My husband squeezed five tablespoons of pus from the swollen lesions.

We had not regularly bathed in diluted bleach, as advised. I tried it a few times. After my skin turned reptilian, I gave it up. But we had devised other, equally elaborate hygiene measures to contain the bug. We washed. We laundered. We maintained a sterile box with hand sanitizers, disposable gauzes, and antiseptic sprays. A set of cast-off pots lived on the stove, for boiling bandages and compresses, which we did religiously.

It didn't matter. Six months after my son's boils had healed, a burning spot appeared on the back of my thigh.

With the help of a small hand mirror and some contortions—at last, I thought, a practical application for years of yoga classes—I could see a small spider bite, one that felt as if a torch were being held to my skin. It soon turned swollen and hard. I stopped wearing jeans, and then pants, to avoid any inadvertent tug or millibar of pressure. Five days passed before I limped into my doctor's office, where she took her scalpel to it and started to dig. Half an hour later, I staggered home in tears, with a giant wad of gauze to soak up the MRSA-infested pus that poured out for days.

MRSA had expressed an ability critical to its effectiveness as a human pathogen: despite our awareness of its presence, and our admittedly half-hearted attempts at controlling it, it had successfully spread from one human body into another. Its basic reproductive number, in our little population, had crossed the critical threshold of 1.

⊙

The toll of pathogens like MRSA, SARS, West Nile virus, and even Ebola are relatively minor in the grander scheme of things. More people die in car accidents in the United States every year than these new pathogens have managed to fell during their collective tenures on Earth. The reason to pay attention to them regardless is that they've begun a journey that pathogens such as cholera completed. And we can see where that road leads.

The transformations that had birthed *Vibrio cholerae* in the wetlands of the Sundarbans were momentous, to be sure. The vibrio had come a long way from its origins as a placid marine bacterium, floating in the sea. But as a pathogen, it still faced an uncertain future. In order to cause a pandemic, a pathogen must infect a large proportion of the human population. The trouble is that human populations are spread

out over wide distances. To do it, cholera would have to become ubiquitous, crossing oceans and traversing continents, snaking thousands of
miles across desert and tundra. And yet pathogens themselves are microscopic and immobile. They have no wings or legs or any other independent means of locomotion. On their own, they are as isolated as island
castaways, marooned in their obscure birthplaces.

To progress to the next stage in the journey to pandemicity, cholera
would have to rely almost entirely upon us.

TWO

LOCOMOTION

I first heard about Chewy, a pet prairie dog, from the microbiologist Mark Slifka outside a banquet room in a Westin hotel in Boston, one chilly November morning. Slifka, one of the world's foremost experts on poxviruses, had just delivered a plenary address to a small group of infectious-disease specialists gathered for a conference about the dynamics of epidemics.

Chewy was a newly acquired pet, Slifka told me, who had been brought to a veterinarian's office by its owners in 2003. They were concerned about their pet's sneezing and coughing. The veterinarian decided to nebulize Chewy with oxygen, encasing him in a hamster ball, a hollow sphere made of plastic, while forcing a jet of oxygen into the ball through a vent.

What the veterinarian didn't know is that Chewy, though a native species (prairie dogs are a kind of North American ground squirrel), had been exposed to killer pathogens from half a world away. The animal had been housed in a pet distribution center alongside a crate of creatures destined for the pet trade that had arrived on a commercial flight from Ghana.[1] That crate included two Gambian giant pouched rats, nine dormice, and three rope squirrels, which had been trapped near the town of Sogakofe, Ghana, where 40 percent of the local rope squirrels, and more than a third of local people, have been exposed to poxviruses.[2]

Little Chewy's sneezes and coughs, along with those of a few dozen other prairie dogs who'd been housed at the same distribution center, were the result of infection with monkeypox. The virus had created pox lesions under Chewy's fur, which seeped the virus, and in his lungs, which spewed the virus in his exhales. Chewy's vet, by nebulizing him in a hamster ball, had just filled the sphere with aerosolized virus. In other words, Slifka said, the vet had created a poxvirus bomb.

When the vet cracked open the ball to let Chewy out, the bomb detonated. A cloud of monkeypox virions wafted into the room. The virus infected ten people who were in the room or who happened to walk through it sometime later. Ultimately, Chewy and other prairie dogs infected by the Ghanaian pets spread monkeypox to seventy-two people in six U.S. states. Luckily, the introduced monkeypox was of the less virulent West African subtype, rather than the more deadly Central African subtype. Only nineteen victims had to be hospitalized.[3]

Slifka liked this story, I think, for the irony of a veterinarian innocently manufacturing a deadly bioweapon. But monkeypox would never have been able to leap out of the jungles of Ghana and into little Chewy's body had it not been for commercial air travel, which gave the pathogen its wings and the free ride it needed to disseminate itself across the globe. The flights that ferried that crate of monkeypox-infested rodents into the United States also brought *Pseudogymnoascus destructans*, a fungus from Europe, into New York. The fungus, which likely hitched a ride from deep inside European bat caves into American ones on the muddy boots of spelunkers, invades and digests the skin of bats. Between 2006 and 2012, the disease it causes, white-nose syndrome, has killed millions of bats in sixteen U.S. states and four Canadian provinces, leading to population declines of up to 80 percent.[4]

Air travel doesn't just ferry new pathogens around; it also dictates the shape and spread of the pandemics they can cause. If you plot a modern pandemic of influenza on a map of the world as the theoretical physicist Dirk Brockmann did in 2013, the pattern will be chaotic and formless. Disease might first erupt in mainland China and Hong Kong, then randomly skip all the way over to Europe and North America with no stops in between, just as a monkeypox outbreak that started in Ghana might next appear in an animal distribution center in Texas. There's seemingly no pattern to explain the spread, or where in the world the pathogen might skip to next.

But Brockmann found that if you track that same pandemic on a map that plots locales in terms of their proximity via air travel, a revealing picture takes shape. On such a map, New York City is closer to London, England, over three thousand miles away, than to Providence, Rhode Island, just three hundred miles away, because of the availability of direct flights. Plotting the spread of a pandemic on a flight-time map does not result in the chaotic eruptions seen on a geographic map. The pandemic resolves into a series of waves, radiating outward one by one like the ripples of a stone dropped into a lake. Our transportation network, Brockmann's map shows, shapes the pandemic more than our physical geography.[5]

⊙

Cholera could never have caused pandemics without the new modes of transport developed in the nineteenth century. On the eve of cholera's debut on the international stage, travel by sea had just started to remake the industrial world, with rapid sailboats and steamers crisscrossing oceans, and newly built canals ferrying people and commodities deep into the interior of nations. We couldn't have devised a better transit system for a waterborne pathogen like *Vibrio cholerae* if we'd tried.

One might think that a marine creature like *Vibrio cholerae*, with access to the ocean, could make its way to almost any shore. The ocean's waters, after all, are connected and in constant circulation. And the second-fastest ocean current in the world—the Agulhas Current—carries water from cholera's home in the southwest Indian Ocean straight to the southern tip of Africa, at the threshold of the Atlantic.[6] Surely a few pioneering vibrio-infested copepods could get swept into its stream and make their way out of South Asia.

But in fact, cholera vibrios, under their own locomotion, are nearly stationary. More than 75 percent of the copepod species they live in and on tend to stay put in the shallow surface waters in which they evolved. The few that might catch a ride on an ocean current are quickly crushed by the deep waters of the open ocean, the marine equivalent of the Sahara, where sustenance is dangerously scarce and progress slow.[7]

Humans can carry the microbe, of course, but only so far. A cholera victim is indeed a walking broadcaster of the bug, spewing the vibrio in his or her stool and on stool-contaminated hands and personal items. But cholera's tenure in the human body is short, about a week at most, provided the vibrio doesn't kill its victim before that. In the nineteenth century, when cholera first emerged, that was hardly enough time to cover the nearly five thousand miles between the Sundarbans and, say, densely populated Europe.

To make the overland crossing, cholera would require large numbers of people moving together. An army of susceptible victims sequentially infected could extend the vibrio's viability over time and miles. This form of travel, from the perspective of the pathogen, would have made for halting progress. If too many people sickened at once, the bug could die out, with all of its potential carriers rendered dead or immune. But at the same time, if only a few fell ill, the pathogen's chances of sequentially infecting enough travelers to carry it over the long haul diminished.

And even that would get the vibrio across only the landmasses of the Old World. To ignite a global pandemic, cholera would have to unlock access to the New World and its bustling nineteenth-century populations of susceptible settlers, slaves, and natives. Cholera would have to cross the deep waters of the ocean. Someone—or something—would have to carry it.

⊙

Europeans and Americans felt that cholera, as a disease of backward Orientals, would never reach the enlightened West. Cholera was an "exotic production . . . developed in the uncultivated, arid plains of Asia," an 1831 French tome declared. They pointedly referred to it as the "Asiatic" cholera to differentiate it from their ordinary diarrhea, which they called "cholera morbus."[8]

France, for example, had little to fear: "In no country but England are the rules of hygiene more faithfully observed," one French commentator proudly opined.[9] Paris, where the wealthy enjoyed airy courtyards and marbled baths in perfumed water, was nothing like the swampy, mangrove-covered Sundarbans.[10] On the contrary, Paris was the center of the Enlightenment. Students of medicine from around the world descended upon the city's new hospitals to learn the latest techniques and discoveries from leading French physicians.[11]

And yet, slowly but surely, cholera arrived on Europe's doorstep. By the fall of 1817, cholera had traveled sixteen hundred miles upstream on the Ganges, killing five thousand at a military camp. By 1824, cholera had radiated into China and Persia, before freezing out in Russia that winter. A second outbreak began in India a few years later. In 1827, British troops invaded Punjab; in 1830, Russian soldiers marched off to occupy Poland. Cholera followed them like a shadow.[12]

Cholera took hold of Paris in late March 1832. Without the benefit of modern medicine, it killed one-half of those whom it infected, causing a set of uniquely horrifying symptoms. There was no tragic tubercular cough or romantic malarial fever. Within hours, cholera's dehydrating effect shriveled victims' faces, wrinkling skin and hollowing cheeks, drying up tear ducts. Fluid blood turned tarry, clotting in the bloodstream. Muscles, deprived of oxygen, shuddered so violently that they sometimes tore. As the organs collapsed in turn, victims fell into acute shock, all the while fully conscious and expelling massive amounts of liquid stool.[13]

Mythic tales circulated of people who sat down to dinner and died by dessert; men who returned home from work to find a note on the door saying that the wife and family lay dying in the hospital; people riding the train suddenly collapsed in front of their fellow passengers.[14] And they did not just clutch their hearts and crumple to the floor, either; their bowels released uncontrollable floods. Cholera was humiliating, uncivilized, an affront to nineteenth-century sensibilities. This exotic invader, the historian Richard Evans writes, transformed enlightened Europeans into a race of savages.[15]

"The thought that one might oneself suddenly be seized with an uncontrollable, massive attack of diarrhoea in a tram, in a restaurant, or on the street, in the presence of scores or hundreds of respectable people," Evans writes, "must have been almost as terrifying as the thought of death itself."[16] Indeed, perhaps more so.

One of the many abiding fears cholera inflamed was of premature burial. Today, we have monitors that beep and buzz when our vital organs fail, and, except for a few headline-grabbing cases, the gray area between life and death is pretty narrow. In the nineteenth century, that gray band was much wider, and stories of exhumed bodies buried in neat shrouds later discovered in contorted positions, bones broken, skeletal hands wrapped in torn-out hair, filled newspapers and magazines, evidence of epic chthonic struggles.

Physicians had been arguing for centuries over the precise symptoms of death, and about the differences between what they called
"apparent" and "real" death. In 1740, the prominent French physician
Jean-Jacques Winslow had argued that some of the common tests for
death—pinpricks and surgical incisions—lacked a certain precision.
(Poor Winslow himself had been mistakenly declared dead and boxed
in a coffin twice as a child.) Some said that the most reliable sign of
death was the putrefaction of the body. But that was a stringent and
stinky test for the bereaved, who might be compelled to wait around for
the decay of their loved one before mourning. And even then, some
argued, the corpse might still be alive, simply comatose and gangrenous.

New laws, inventions, and methods of coping with dead, or apparently dead, bodies helped diminish the problem. In the 1790s, a new
system implemented at Paris mortuaries required that corpses be outfitted with special gloves, such that if a corpse's finger so much as trembled,
a string would be pulled and a large hammer would slam down on an
alarm. Guards patrolled the mortuary under the direction of local physicians, ears peeled. (Today, we surveil the living for signs of death; back
then, they surveilled the dead for signs of life.) An 1803 law required a
day's delay between an apparent death and the subsequent burial, just
in case someone got it wrong. In 1819, the French physician René-
Théophile-Hyacinthe Laënnec developed the stethoscope, which made
audible even the sound of a faintly beating heart (simultaneously freeing
gallant physicians from suggestively pressing their ears to their female
patients' chests). Charity groups such as the Royal Humane Society
formed, for the express purpose of resuscitating the drowned, launching
public awareness campaigns on the finer distinctions between the living and the dead. (Their motto, which they've kept to this day, is *Lateat
scintillula forsan*, "A small spark may perhaps lie hid.")[17]

Cholera terrified Parisians by destroying these few safeguards. Cholera
could easily make a living person look like a corpse: blue, sunken, still.
"It is so easy to be completely mistaken," one physician complained during the 1832 cholera outbreak, "that I once marked down as dead an
individual who in fact died only several hours later."[18] And yet during
the epidemic, burial-delay rules were overturned. Dead bodies—along
with apparently dead bodies—were piled onto rickety wagons like so
much freight, occasionally spilling their contents onto the streets. All
were summarily buried in mass graves, layered three bodies thick.

Local authorities outlawed public gatherings and banned markets within the city center. They marked victims' homes, quarantining survivors inside. Still, the funeral parades continued. The churches were hung in black. The city's hospitals overflowed with perfectly still patients on the edge of death, their skin turned shockingly purple by cholera's depredations. Still-alive cholera patients anesthetized themselves with alcoholic punch passed off as medicine. "It was a fiendish sight," the visiting American journalist N. P. Willis wrote. "They were sitting up, and reaching from one bed to the other, and with their still pallid faces and blue lips, and the hospital dress of white, they looked like so many carousing corpses." Spattered blood and bodily fluids leaked from wagons carrying the dead along the city's cobblestone roads.

In the evenings during that terrible spring, Paris's elite attended elaborate masquerade parties where, in denial and defiance of cholera's toll, they danced to "cholera waltzes," costumed as the ghoulish corpses many would soon become. Willis, who attended one of the so-called cholera balls, described a man dressed as cholera itself, with "skeleton armor, bloodshot eyes, and other horrible appurtenances of a walking pestilence." Every now and then, one of the revelers would rip off his mask, face purpled, and collapse. Cholera killed them so fast they went to their graves still clothed in their costumes.[19] (Paris's cholera balls, and Willis's reporting on them, inspired a mordant thirty-three-year-old writer in Baltimore—Edgar Allan Poe—to pen "Masque of the Red Death," a short story about a masquerade ball in which the entry of a masked figure "shrouded from head to foot in the habiliments of the grave" leads to the death of "revellers in the blood-bedewed halls of their revel.")

By the middle of April, cholera had killed more than seven thousand Parisians. The final death count remains obscure. To reduce panic, the government stopped publishing death statistics altogether.[20]

Those who could fled the city, leaving behind a collapsed society in which cholera could rage undeterred by the ministrations of nurses, doctors, or police officers.[21] "Cholera! Cholera! It is now the only topic," bemoaned Willis. "People walk the streets with camphor bags and vinaigrettes at their nostrils—there is a universal terror in all classes, and a general flight of all who can afford to get away." Some fifty thousand panicked Parisians decamped, a swarming exodus that spilled over the roads, rivers, and seas, carrying cholera to new lands even more efficiently than the hordes of sailors, traders, and soldiers had before them.[22]

They escaped on foot, they boarded carriages, they paddled down-stream, and they boarded oceangoing vessels. Thanks to new trade routes, they quickly brought cholera across the sea and deep into the interior of North America.

⊙

For centuries after Columbus, crossing the Atlantic had been risky and only sporadically attempted. The Dutch, who settled what would become New York City, chartered ships to cross the ocean only once a year, if that. The difficult and costly passage took eight weeks, in part because cautious captains steered clear of the shortest route over the forbidding North Atlantic. During the British colonial era, restrictive tariffs and the predations of pirates strangled the aspirations of those shipowners who hoped to ferry goods and people across the Atlantic, quieting the ports of New York, Boston, and Philadelphia. Even after Americans gained independence from Britain, the only way to get oneself across the Atlantic was to wait for a local shipowner to advertise a date of departure and then hope that sufficient cargo and other passengers would sign up, too, and then, if all of those stars aligned, languish in a port city, often for a week or more, for the wind and weather to cooperate.

The shipping trade out of the United States started to pick up during the Napoleonic Wars, as the ports of New York, Boston, and Philadel-phia captured a slice of the lucrative seagoing trade with China while Europe roiled. In 1817, just as cholera emerged in the Sundarbans, am-bitious American shipowners, financed by the newly established Bank of the Manhattan Company (later to become the multinational behemoth JPMorgan Chase), established something novel in transatlantic ship-ping: regular scheduled service between American ports and those in Europe, including Liverpool, London, and Le Havre. No more waiting around at the docks. The sailing packets, as they were called—the Black Ball Line, the Cunard Line, and others—set sail weekly from the United States, loaded with passengers, sacks of mail, and other commodities to ferry to and fro across the ocean.[23]

During all of the seventeenth and eighteenth centuries, only some four hundred thousand emigrants from Europe had made it to the New World. In less than a century after the introduction of the transatlantic sailing packets, 30 million Europeans boarded ships bound for the

United States. The Atlantic, once a formidable ecological barrier to cholera's spread, had become a veritable thoroughfare for people, cargo, and the invisible microbes they unknowingly carried with them.

⊙

Infected passengers aboard the packets easily passed the vibrio to the uninfected. While first-class passengers enjoyed stylish quarters and elaborate meals, most passengers at sea traveled cheek by jowl in steerage, their unwashed hands and bodies pressed tight together. At night and during bad weather, the hatches had to be closed, trapping steerage passengers belowdecks in the dank, heavy air. "How can a steerage passenger remember that he is a human being when he must first pick the worms from his food," one journalist who made the journey complained, "and eat in his stuffy, stinking bunk, or in the hot and fetid atmosphere of a compartment where 150 men sleep?" Several hundred passengers might have access to just a handful of toilets, and excreta, mingled with stinking bilge water, trickled through the decks.[24]

Shipboard practices themselves could introduce cholera to passengers. Before setting sail, ships often filled their drinking-water casks from the same streams and bays in which local people bathed and defecated. If cholera had struck any of the port towns and cities from which ships departed or visited en route, a few local vibrios could easily be sucked up into the drinking-water supplies aboard. This water was then carried across the sea in wooden casks and tanks that were rarely, if ever, cleaned. Transatlantic passengers both drank and cooked with that water during the journey.[25]

Once cholera struck the passengers, the entire ship became a roving disseminator of cholera vibrio, depositing contaminated excreta directly into the seas, bays, and harbors through which they passed.[26]

The ships could by themselves ferry cholera vibrio, too, even if passengers aboard remained unscathed. Nineteenth-century ships carried all manner of mammals, birds, plants, and other creatures as witting and unwitting cargo. Livestock, companion animals, and pests skittered aboard. Barnacles, mollusks, algae, and other marine creatures vulnerable to colonization by cholera vibrio bored into and attached themselves to ships' wooden hulls, making possible passages that would never have been accomplished of their own accord. (*Sphaeroma terebrans*, a tiny crustacean that burrows into the root tips of mangrove trees, was one such hanger-on. It bored into a wooden ship hull sometime in the 1870s and made its way from its Indian Ocean home to the Atlantic, where it now abounds, munching on the roots of mangrove trees in Florida and elsewhere.)

Ships spread thousands of species around the world in their ballast, heavy materials used to fill an empty ship's holds to keep it stable in the water. Wooden vessels used dry ballast—tons of sand, soil, and stone— alive with crab, shrimp, jellyfish, sea anemones, sea grasses, and algae, among other creatures. They'd shovel it aboard when they left shore and dump it miles later upon reaching their destinations, creating massive deposits full of alien invaders. A few cholera-vibrio-infested crustaceans, dumped overboard in a pile of dry ballast, could seed a new immigrant colony across an ocean.

Water ballast, used in iron ships, carried cholera even more efficiently. Iron ships, besides being watertight and making water ballast possible, were faster and stronger, and had more room for storage than wooden ships. The first iron steamboat was built in 1820 and traveled from London to Le Havre, France, and then to Paris. By 1832, Europe's iron ships steamed to Africa and India.

As a method of transporting marine organisms, writes the marine ecologist J. T. Carlton, ballast water "appears to have few if any parallels on land or at sea for its biological breadth and efficiency."[27] Modern studies suggest that ballast water carries some fifteen thousand marine species across the oceans and seas every week, cholera vibrio among them. Each one of the millions of gallons of ballast water sucked up from the shallow bays and estuaries of cholera-plagued Europe and Asia could hold some tens of billions of viruslike particles, awaiting their release across the sea.[28]

⊙

By land, the interior of the United States at the time of cholera's emergence was largely impenetrable. Most of the nation's roads were little more than muddy tracks through wild forests and swamps, on which fallen trees and mud could strand horse-drawn carriages and carts for weeks. Transporting goods only a few dozen miles into and out of the interior of the country by land required as much time and money as shipping them across the ocean to England.[29]

Boats, in contrast, moved swiftly and reliably. Newly developed steamboats allowed passengers to travel up and down natural waterways such as the three-hundred-mile-long Hudson River, which ran from the Adirondack Mountains to New York City, and the two-thousand-mile Mississippi River, which flowed from northern Minnesota to the Gulf of Mexico.

But before the mid-1800s, the Appalachian Mountains' unbroken spine of peaks down the eastern half of the country served as a gargantuan wall separating the shipping trade along the Mississippi River and the Great Lakes from the international oceangoing trade along the Hudson and the Atlantic Ocean.[30] Cholera or any other waterborne pathogen that made it to U.S. shores was blocked from the waterways farther west.

The Erie Canal, which opened in 1825, changed all that, wedding the salt water of the Atlantic with the freshwater web of the interior. The canal cut right through the Appalachians, connecting the Hudson River to Lake Erie, more than three hundred miles away, at Buffalo. It was a marvel of engineering, bought for the then astronomical sum of $7 million (that's some $130 billion in 2010 dollars), a "watery highway," as Nathaniel Hawthorne put it, "crowded with the commerce of two worlds, till then inaccessible to each other." By slashing transportation costs between the interior of the country and its coasts by 95 percent, the canal transformed the economy of the port at its southern terminus, New York City. Thanks to the canal, New York would eclipse its rival port cities, Philadelphia, Boston, and Charleston, becoming "a city of countless ships, which line both the banks to a considerable distance, with a forest of masts, to which few other cities can present a parallel," as one observer put it.[31]

But while the canal dramatically boosted commerce, it also allowed the microbial pathogens of the rest of the world to wash into every corner of American society. To celebrate the opening of the canal, dignitaries had poured water bottled from thirteen of the world's great rivers—the Ganges, the Nile, the Thames, the Seine, the Amazon, and others—into the swirling waters of New York harbor, along with a keg of water from the canal itself. They were celebrating the new ease of waterborne trade, but their ceremony more accurately evoked the new era of waterborne disease they'd begun.[32]

Traffic on the canal was intense. Narrow canal boats departed daily from even the tiniest villages along the way, running all day and night.

Thirty thousand people toiled on the canal's eighty-three locks and aqueducts and guided the horses and mules that pulled boats through its
waters, whole families living along its path to ensure its daily function-
ing. By 1832, half a million barrels of flour and more than one hundred
thousand bushels of wheat—not to mention 36 million feet of timber
that year alone—had been sent down the Erie's stagnant, shallow
waters. Canal boats, piled high with logs and crammed with passengers, sometimes had to wait for as long as thirty-six hours in queues at
the locks.

And along with the wheat and tea came waves of immigrants. They
disembarked from schooners that crossed the Atlantic, rode along the
canal, and transferred to new vessels for the continuing journey westward
on the water, bringing cholera with them.[33]

⊙

In the spring of 1832, tens of thousands of immigrants from cholera-
plagued Europe arrived at seaports along the eastern coast of North
America. Cholera first struck in Montreal and Quebec City, at the northeastern terminus of the spider's web of rivers and canals that sprawled
across North America. Over eleven brutal days, cholera killed three
thousand in those two Canadian cities and started breaking out in
nearby canal towns. Once cholera made it into the canal system, it had
secured its ticket into the rest of the continent. Scores of soldiers from
New York were heading west to fight the Native American warrior Black
Hawk over disputed territory in Illinois. Cholera followed them westward like a shadow. Dozens of soldiers fell ill on the riverboats and were
left behind along the way, seeding new outbreaks. Others, terrified, deserted. A passerby encountered six cholera-plagued deserters along the
road between Detroit and Fort Gratiot, Michigan, at the southern tip of
Lake Huron; hogs consumed the corpse of the seventh. "Some died in
the woods and were devoured by wolves," writes the cholera historian
J. S. Chambers. "Others fell in the fields and along the roadside but were
left untouched where they fell. The straggling survivors, with their knap-
sacks on their backs, and shunned as the source of a mortal pestilence,
wandered they knew not whither." Of the entire contingent, more than
half died or deserted "without ever firing a shot."

Downstream in New York City, more than seventy thousand residents panicked by the news of cholera's arrival in North America fled.[34]

☉

Very little of the grand canal era that the Erie Canal ushered in remains today. The current state of the C&O Canal in Maryland testifies to its precipitous decline. The canal, which ferried coal from the Allegheny Mountains from 1831 to 1924, is now primarily used as a recreational area. The long ditch is mostly dry, and the old lockhouses, where lock-keepers and their families once lived, are in ruins. Only their stone foundations and their nearby water pumps remain, hidden behind scrubby pawpaw trees. Their outhouses, situated a handful of yards away, have been replaced by light blue porta potties for passing bicyclists outfitted in garish synthetic fabrics, zooming down the canal's old towpath where mules and horses once pulled boats. The draw here, once again, is the wild, shallow, and barely navigable river, which attracts kayakers and canoeists, and the odd local kid crashing through the woods for a quick summer's dip.[35]

But even as the canals fade into oblivion, the avalanche of trade and mobility that they began continues, gaining speed.

Canals and steam engines, along with coal and cotton gins and other miracles of the factory age, were the first to pry the global economy out of its historical stranglehold. For hundreds of years, world economic production had remained relatively flat, rising just 1.7 percent per capita per century, as barely fed humans scratched out a living on the meager strength of their own metabolism. Then we unleashed the buried energy of fossil fuels, sparking the Industrial Revolution. In less than a century—between 1820 and 1900—world economic production doubled. It's continued to expand ever since. In the last sixty years, global trade has increased a whopping twentyfold, faster than population or GDP growth.[36]

Canals sowed the seeds of their own demise. By introducing Americans to the world of international commerce—for the first time, farmers in Buffalo could enjoy fresh Long Island oysters and exotic foreign commodities like tea and sugar—they sparked an appetite that they'd never be able to satisfy. Demand for ever faster, more powerful transportation grew like a cancer and the canals could not hope to keep up. They were, after all, only four feet deep. First came the railroads. Then the highways. Finally, today, the airplane—which carries global trade's most high-value products—has eclipsed them all.

The machine invented by the Wright brothers in 1903 now carries one billion human beings through the clouds every year.[37] They don't just fly in and out of a handful of prominent airports in major cities, but into and out of tens of thousands of airports in small towns and minor cities in even the most remote and far-flung nations. There are some fifteen thousand airports in the United States, but not only that: there are also more than two hundred in the Democratic Republic of Congo, one hundred in Thailand, and, as of 2013, nearly five hundred in China.[38]

New York City is no longer the center of today's global transportation network, of course. The hub has shifted. Of the ten largest and busiest airports in the world, nine are in Asia, seven in China alone.[39] And just as the United States' gateway to the world was once New York City, China's gateway to the world is Hong Kong, where more cargo— both visible and invisible—is loaded onto airplanes than anywhere else. Whereas cholera sailed and steamed around the world, cholera's children fly.[40]

⊙

The growth of wet markets created the conditions for the SARS virus to spill over and adapt to humans, but it was the modern air travel network and a single establishment—a nondescript business hotel called the Metropole in the middle of Kowloon in Hong Kong—that distributed it across the planet, triggering the global outbreak of 2003.

SARS's first victims in south China had been rushed to local hospitals, including the Sun Yat-Sen Memorial Hospital in Guangzhou. There, clinicians working around the clock provided whatever care they could, but they also continued living their lives. One, Dr. Liu Jianlun, finished his shift tending SARS patients, then cleaned up, changed clothes, and left Guangzhou for the ninety-mile trip south to Hong Kong to attend a wedding. A few hours later, he checked into Room 911 at the Metropole, which is where the SARS virions in his body made their escape.[41]

So much virus took leave of his body in that room that investigators recovered genetic evidence of the virus in the carpet months later.[42] Just how SARS spread from Dr. Liu to twelve other hotel residents remains unclear. Perhaps they shared an elevator ride with him or passed through the hallway outside his room after he'd coughed or vomited. Or they touched the corridor walls after he had brushed against them with a hand he'd sneezed into. Or inhaled some of the aerosolized virus that had escaped from his room after he flushed the toilet.[43]

What we do know is that Liu's fellow hotel residents were a mobile, international bunch. So were mine, when I visited the hotel (now called the Metropark) in the winter of 2012. In the dimly lit hotel bar, outfitted with a drop ceiling covered in glossy black tiles, Spanish-speaking couples quietly downed rounds of shots while a white-haired Australian browsed the business section of an English-language newspaper. A little later, I overheard him discussing his financial dealings in Tanzania and Indonesia with a trim Asian businessman.

One of Liu's fellow residents in 2003 was a flight attendant. She made it as far as Singapore before being hospitalized, where she passed on the virus to her doctor, who planned to fly to New York, where he was to attend a medical conference. He made it as far as Frankfurt, Germany. Others exposed to Liu at the Metropole boarded planes to Singapore, Vietnam, Canada, Ireland, and the United States. Within twenty-four hours, the SARS virus from Liu had spread to five countries; ultimately, SARS appeared in thirty-two countries. Thanks to the miracle of air travel, one infected man seeded a global outbreak.[44]

⊙

Many people worry about catching bugs during air travel, but in fact only a subset of pathogens easily spreads during flights themselves. Pathogens that spread via direct contact, like HIV and Ebola, are unlikely to amplify during flights. Only two Ebola-infected people are known to have traveled by air during the first year of the 2014 epidemic in West Africa, and neither of them infected anyone else on the flights themselves.[45] (Contact-transmission pathogens like Ebola are much better suited to take advantage of practices like burial rituals, in which people ceremonially bathe infected corpses, and health-care settings in which clinicians extensively handle infected patients, both of which played important roles in fueling the 2014 Ebola outbreak.) Pathogens that spread between people through vectors, like mosquitoborne West Nile virus and dengue, can only occasionally survive air travel, too—the cool, arid atmosphere of modern airplanes is often deadly for their mosquito carriers.

Respiratory pathogens like SARS, however, are ideally suited. By spreading through droplets released while coughing or sneezing or through aerosols, extra-tiny droplets that can hang suspended in the air, they can turn a single infected carrier upon departure into a planeload of carriers upon arrival. A less noted but equally potent way air travel broadcasts pathogens is by extending the mobility of infected carriers who would have been far too fragile to travel via other means. Surgical patients, for example, played little role in the global spread of infectious pathogens in the past. People who'd recently undergone surgery were relatively immobile. Not so today. Surgical patients travel the world, carrying pathogens from operating rooms on one side of the globe to the other.

Every year, for example, hundreds of thousands of so-called medical tourists from the United States, Europe, the Middle East, and elsewhere fly to countries such as India to undergo surgery. Thanks to market reforms in the early 1990s, which unleashed several decades of 8 percent annual growth in the Indian economy, modern private hospitals in India now provide the same standard of care as Western hospitals. But because poverty and low wages persist in the country (among other reasons), they can do so at a fraction of the cost. As a result, foreign patients looking for an affordable organ transplant, or a knee replacement, or heart surgery, arrive in droves.[46]

This is a striking reversal from as recently as the 1980s, when India was still an economic backwater, and families like mine who went there to visit relatives packed their suitcases full of any medical supplies we thought we might need so we wouldn't have to rely on spotty local health services. Locals who could afford it flew to New York or London for high-tech medical attention.

Back then, airports in Indian cities were crumbling fluorescent-lit buildings where gangs of skinny, mustached young men in tight buttondown shirts hawked taxi rides, while knots of bedraggled multigenerational families clutched their tickets anxiously. One hoped not to have to use the alarmingly blocked and overflowing public toilets. Today, the New Delhi Indira Gandhi International Airport is a sparkling facility, complete with upscale cafés, gigantic, colorful abstract murals, and moving walkways ferrying trendy young business travelers and their ubiquitous tiny black gadgets. When I visited in 2012, I found signposted directions to Medanta Hospital, one of many gleaming new corporate-owned private hospitals that cater to medical tourists, conveniently located right outside the baggage claim.[47]

Fifteen percent of Medanta's patients arrive from overseas for surgery, which costs them one-fifth the price they'd pay in Western countries, hospital spokespeople say. The hospital itself, a short drive from the airport, is an impressive building with expansive verdant gardens, encircled by tall wrought-iron gates that seal it from the old world that exists just beyond: hawkers selling freshly squeezed juices from fly-covered wooden carts, workers crouched on their haunches cooking over smoky fires. Inside, the hospital looks and feels like a museum, with soaring ceilings, white marble tiled floors, and giant walls of frosted glass.

Behind one frosted glass door is a special lounge reserved for medical tourists and their relatives. The only Indians here are behind the counters: the others are of East Asian, Middle Eastern, and Western ancestry, still toting their giant shrink-wrapped suitcases. They loll on black leather couches and enjoy free hot drinks while watching flat-screen televisions. The hospital's International Patient Services team organizes their treatment packages, assists them with their visas, provides airport pickup, and arranges hotel reservations and postrecovery sightseeing trips. They even provide a concierge service for dining and entertainment.[48]

But as comfortable as these medical tourists may be, once under the knife their internal tissues will be exposed to New Delhi's unique microbial environment, and they bring any microbes they gain during their procedures back home. People who undergo surgeries are especially vulnerable to infectious pathogens. Surgeons' knives breach the protective layer of skin that separates the interior of the body from the exterior environment, allowing the army of microbes that live on the surface of the skin, in the air above the bed, and on the surgical instruments and other objects that pass over the open wound to slip in. Even the most elaborate sterilization practices often fail to stop them. Those that make their way into the body are likely to flourish, since the surgery itself, not to mention the conditions that often precede it, depress patients' immune systems.

Hospitals like Medanta boast infection rates as good as or lower than those of hospitals in the United States, but the bacteria responsible for these infections are not the same. For one thing, most bacteria in Indian hospitals are gram-negative, which means they are encased in tough outer membranes that make them more resistant to antibiotics and antiseptics than the gram-positive strains that dominate in Western hospitals. (The term gets its name from Hans Christian Gram, the developer of the test that distinguishes between the two types.) For another thing, since India suffers a heavy burden from bacterial diseases—diarrhea and tuberculosis kill around a million people every year—and does not regulate the use of antibiotics (they're widely available across the country without a prescription), many of India's bacterial pathogens are impervious to antibiotics. Compared to about 20 percent of hospital infections in the United States, more than half of hospital infections in India are resistant to common antibiotics.[49]

A particularly pernicious pathogen called New Delhi metallo-beta-lactamase 1 (NDM-1) has been present in New Delhi since at least 2006. It's actually a fragment of DNA called a plasmid, which can spread between bacterial species. What makes it dangerous is that it endows bacteria with the ability to resist fourteen classes of antibiotics, including the powerful intravenous antibiotics administered solely in hospitals as a last resort in patients who have failed to respond to all other treatment options. When NDM-1 inserts itself into a bacterial pathogen, in other words, it makes that strain nearly untreatable. Only two imperfect drugs can contain NDM-1 infections: an older antibiotic called colistin, which fell into disuse in the 1980s because of its toxicity, and an expensive IV antibiotic called tigecycline, which is currently approved only for soft-tissue infections.[50]

Thanks to the power, speed, and relative comfort of air travel, even the most obscure pathogens can leap over continents and oceans. NDM-1 escaped Indian operating rooms in the bodies of medical tourists. In 2008, during a routine test to measure bacteria levels, NDM-1 bacteria were isolated from the urine of a fifty-nine-year-old man hospitalized outside Stockholm. The man had acquired the bacteria in New Delhi. Other cases appeared in Sweden and also in the U.K., all connected to patients who'd traveled to India or Pakistan for procedures such as cosmetic surgeries or organ transplants. In 2010, three patients in the United States were found to be infected with NDM-1 isolates; all three had received medical care in India. By 2011, NDM-1 bacteria had been isolated from patients in Turkey, Spain, Ireland, and the Czech Republic. By 2012, medical tourists had helped NDM-1 radiate into twenty-nine countries around the world.[51]

So far, NDM-1 has mostly been found in bacterial species that can live harmlessly inside the body, like the *Klebsiella pneumoniae* that resides in healthy people's mouth, skin, and intestines, and *Escherichia coli*, which can be found in their guts. But the medical tourism industry that has helped disseminate the plasmid remains lucrative and robust. Health-care costs continue to skyrocket in industrialized countries, forcing patients out of their homes and into the air in search of cheaper, faster treatments. Despite the emergence of NDM-1, they show no signs of wanting to change their tickets. The farther they and other carriers take NDM-1—and the more bacterial species it encounters in its peregrinations—the more likely the plasmid's transfer into a dangerous bacterial pathogen becomes.

Such a pathogen, endowed with NDM-1, would place a ruinous burden on the practice of medicine, causing nearly unstoppable infections. Few medical procedures would be worth the risk. "All medical feats will come to a stop," predicts the medical microbiologist Chand Wattal of Sir Ganga Ram Hospital in New Delhi. "Bone marrow transplants, or this or that replacement—all that will vanish," he says.[52]

⊙

The ease with which our transportation networks dispatch pandemic-worthy pathogens like NDM-1 is discomfiting, and since cholera's time they've done it with increasing speed and efficiency.

But we are not passive victims of our mobility, doomed to be followed by a cloud of malevolent microbial hangers-on. Global distribution is a prerequisite of pandemics, but it's not a sufficient condition on its own. Pathogens, even if ubiquitous, can cause pandemics only if they encounter the right transmission opportunities wherever they land. A widely distributed pathogen, deprived of such opportunities, is as harmless as a defanged snake.

And pathogens' reliance on specific modes of transmission is not particularly flexible. Once adapted to a certain mode of transmission, pathogens cannot easily alter the complex machinery they've evolved in order to jump from victim to victim. That's why, historically, mosquitoborne pathogens don't evolve to become waterborne ones, and waterborne ones don't evolve to become airborne ones. But while their modes of transmission are relatively fixed, the transmission opportunities they exploit are fluid: they're shaped almost entirely by our behavior.

It's true that some pathogens spread by taking advantage of forms of human intimacy that are integral to our societies, such as sexual relations or the proximity that results in people breathing on each other, but many others spread via more obscure, convoluted practices that are comparatively rare, or easily altered. The pathogen *Toxoplasma gondii* spreads into humans when rodents ingest its eggs, cats consume the rodents, and humans then expose themselves to the infected cats' litter boxes. Transmission of the pathogen *Dicrocoelium dendriticum* requires that snails incubate its eggs, ants drink the snails' slime, and then grazing animals feed on the ants.

A pathogen such as *Vibrio cholerae* requires that humans regularly consume their own excreta. That's good news because it means we can easily deprive it of transmission opportunities, for consuming each other's waste is required for neither human survival nor the stability of our societies. The bad news is that sometimes historical conditions conspire to make even the most unnecessary and risky behaviors nearly inevitable.

THREE

FILTH

For pathogens, excreta is a perfect vehicle for spreading from one person to another. Human feces, freshly emerged from the body, teems with bacteria and viruses. By weight, nearly 10 percent is composed of bacteria, and in each gram there might be up to one billion viral particles. Every year, a typical human produces 13 gallons of the stuff (plus 130 gallons of sterile urine), creating a microbe-rich river of waste that, unless contained and isolated, can easily stick to the bottoms of feet, cling to hands, pollute food, and seep into the drinking water, allowing pathogens to creep from one victim to the next.[1]

Fortunately, people have known for centuries that healthful living requires separating ourselves from our waste. Ancient civilizations in Rome, the Indus valley, and the Nile valley knew how to manage their waste so that it didn't contaminate their food and water.[2]

Ancient Romans used water to flush waste far from their settlements, where it could rot undisturbed. The Romans controlled a supply of fresh water from distant, unpopulated highlands via a network of wood and lead pipes, which brought the typical resident three hundred gallons of fresh water every day, three times more water than the average water-guzzling American uses today, according to the Environmental Protection Agency. The Romans mostly used this flow of water to run bathhouses and public fountains, but they also used it in communal latrines, where they sat over keyhole-shaped openings on benches situated over large drains, a gutter of fresh running water flowing at their feet.[3]

One of the principal virtues of using water to flush away excreta, from a public-health perspective, is that during that critical period between production and decomposition, no human need handle the microbe-rich dung. The water simply carries it away. The drawback is that it also makes the excreta mobile, creating a large volume of flowing, contaminated water that can pollute drinking-water supplies (among other things). But for precisely the same reasons that drove the ancients to construct their water distribution networks in the first place—their love of fresh, cleansing waters—the Romans understood the importance of clean drinking water. They turned their noses up at people foolish enough to bathe in, let alone drink, water that wasn't filtered, and heeded the advice of the ancient Greek physician Hippocrates that water be boiled before drinking.[4]

By all rights, these healthful practices should have persisted throughout the ages. But they did not. By the nineteenth century, the European descendants of the ancient Romans who came to populate the city of New York had forsaken the practices of their ancestors. They immersed themselves so completely in each other's waste that each likely ingested two teaspoons of fecal matter every day with their food and drink.[5]

Partly, this about-face had to do with the rise of Christianity in the fourth century A.D. The Greeks and Romans, not to mention the Hindus, the Buddhists, and the Muslims, all prescribed ritualized hygiene practices. Hindus must wash after any number of acts considered "unclean," as well as before prayer. Muslims must perform ablutions at least three times before their five-times-daily prayers, as well as on numerous other occasions. Jews were enjoined to wash before and after each meal, before praying, and after relieving themselves. In contrast, Christianity prescribed no elaborate water-based hygiene rituals. Good Christians had only to sprinkle some holy water to consecrate their bread and wine. Jesus himself, after all, had sat down to eat without washing first. Prominent Christians openly repudiated water's cleansing effect as superficial, vain, and decadent. "A clean body and a clean dress," opined one, "means an unclean soul." The most holy Christians, with their lice-infested hair shirts, were among the least washed people on Earth. Not surprisingly, after the Goths disabled the Roman aqueducts in 537, the unwashed leaders of Christian Europe didn't bother rebuilding them, or any other elaborate water-delivery system.[6]

Then, in the mid-fourteenth century, bubonic plague arrived in Europe. Leaders of Christian Europe, like political leaders everywhere facing an incomprehensible threat, blamed their favorite bugbear, water-based hygiene. In 1348, physicians from the University of Paris condemned hot baths in particular, asserting that bathing with water opened the skin's pores and allowed disease to enter the body. "Steam-baths and bath-houses should be forbidden," King Henry III's surgeon Ambroise Paré agreed. "When one emerges, the flesh and the whole disposition of the body are softened and the pores open, and as a result, pestiferous vapour can rapidly enter the body and cause sudden death," he wrote in 1568. Across the Continent, the remaining bathhouses from the Roman era were shuttered.[7]

Given their suspicions about the moral and mortal dangers posed by water, medieval Europeans used as little as possible to manage their waste and quench their thirst. They drank directly from shallow wells, muddy springs, and stagnant rivers. If it tasted bad, they'd turn their sparse water supplies into beer.[8] Those who could afford it practiced "dry" hygiene. Seventeenth-century aristocratic Europeans masked the ripe odors of their grimy bodies with perfumes and by wrapping themselves in velvets, silks, and linen. "Our usage of linen," asserted a seventeenth-century Parisian architect, "serves to keep the body clean more conveniently than the baths and vapour baths of the ancients could do." They used golden ear picks encrusted with rubies to extract wax from their ears and rubbed their teeth with black silk edged in lace: anything other than wash themselves with water. "Water was the enemy," writes the hygiene historian Katherine Ashenburg, "to be avoided at all costs."[9]

The result was centuries of close congress with human and animal waste, inuring preindustrial people to its presence and even leading them to see it as salubrious. Medieval Europeans commonly lived with the smelly presence of all manner of dung underfoot, their own being the least of it. They shared their homes with the animals that fed and transported them, and the cows, horses, and hogs produced far more prodigious quantities of manure than the local humans did, and were even less fastidious about where to deposit it.[10] To dispose of their own waste, some perched on simple buckets, in their homes or in outhouses, which they called "privies." Slightly more elaborate setups included hand-dug pits either outdoors or in cellars, sometimes loosely lined with stones or bricks (as in cesspools and privy vaults), fitted perhaps with a bottomless seat or a squatting plate. The precise method of collection and disposal depended upon the whim of individual domiciles; political authorities imposed few if any rules.[11] The act of excretion itself didn't require privacy or provoke shame back then as it does now. Sixteenth- and seventeenth-century monarchs such as England's Elizabeth I and France's Louis XIV openly relieved themselves while holding court.[12]

Far from reviling human feces, medieval Europeans even began to think of it as medicinal. According to a history of sanitation by the journalist Rose George, the sixteenth-century German monk Martin Luther ate a spoonful of his own feces every day. Eighteenth-century French courtiers took a different route, ingesting their "poudrette," dried and powdered human feces, by sniffing it up their noses.[13] (Was this dangerous? Quite possibly. But in contrast to more immediate threats like, say, bubonic plague, the sporadic cases of diarrhea these practices may have caused would have paled.)

When Dutch colonists established the little town of New Amsterdam on the southern tip of the island of Manhattan in 1625, they brought these medieval ideas about and methods of sanitation with them. The Dutch built their privies to open at ground level and pour their contents into the streets, so that "hogs may consume the filth and wallow in it," as a New Amsterdam official put it in 1658. The English, who took control of the colony in 1664 and renamed it "New York," similarly contained their excreta in what they called "ordure tubs," which they, too, emptied into the streets.[14]

These medieval practices persisted through the nineteenth century, even as the town of a few thousand inhabitants mushroomed into a small city of several hundred thousand residents. By 1820, privies and cesspools covered one-twelfth of the city, and tens of thousands of hogs, cows, horses, and stray dogs and cats roamed the streets, defecating at will.[15] New York's outhouses and privies "were in a most filthy and disgusting condition," complained one official in 1859, with "accumulations of stagnant fluid, full of all sorts of putrefying matter, the effluvia from which is intolerable." Raw sewage rotted in the back lots and sidewalks of tenement buildings for weeks and months at a time. To cover up the filth, landlords laid wooden boards over the ground. When pressed, the boards exuded a "thick greenish fluid," the city inspector reported.[16]

Occasionally the city government hired private outfits to collect the manure and human waste that built up on the streets. It was sold as fertilizer, a trade that turned Brooklyn and Queens into two of the most productive agricultural counties in mid-nineteenth-century America. But "sewage farming," as it was called, never gained momentum. There was nowhere sufficiently isolated to store the excreta as it awaited transport. The reeking piles left at the wharves provoked complaints from nearby residents. Plus, city authorities tended to dole out the job to private contractors as political patronage, many of whom didn't bother actually doing the job.[17]

As a result, most of the city's excreta simply seeped along the streets and soaked into the ground. The filth compacted into "long ridges forming embankments along the outer edge of the sidewalks," as the newspaper editor Asa Greene put it in the late 1840s.[18] Horses and pedestrians trod upon it, slowly flattening it into a dense mat. The paving stones underneath the thick sludge carpeting the streets "rarely had shown themselves again to mortal eyes," Greene noted in his diary. On the rare occasion that the city scraped the streets clear, locals professed shock. Greene quotes an elderly woman, who'd lived her whole life in the city, remarking on the state of recently cleaned streets: "Where in the world did all these stones come from? I never knew that the streets were covered with stones before. How very droll!"[19]

The use of medieval sanitation methods in early industrial cities created conditions ripe for a cholera epidemic. These places were nothing like the European countryside in which their waste-management habits had formed. In the mostly rural communities that medieval Europeans had lived in, soils were thick and population density low. When their pits of excreta reached capacity, people had the space to simply seal them and dig new pits nearby. The streets that they emptied their chamber pots into were not heavily traveled. The excreta could disperse into the ground, where the soil's disparate particles of minerals, organic matter, and microbes would trap and filter it, allowing it to decompose well before it reached the groundwater.[20]

The island of Manhattan, in contrast, had limited capacity to hold and filter the waste. Manhattan was the largest of a series of islands, now known as Staten, Governors, Liberty, Ellis, Roosevelt, Ward, and Randalls, that dotted the Hudson Estuary. Two brackish rivers flanked the island, pulsing with the Atlantic's tides: the Hudson on its west side and the East River on its east side. The two flows collided just off the island's southern tip, stirring up bottom sediments and sending up plumes of nutrients into the water column. Oysters grew so large they had to be cut into three to eat. (Today, if you dig deep enough almost anywhere in lower Manhattan, you'll hit pure shell, the remains of earlier oyster feasts.) But while local waters were rich with marine life, the island's soil was only three feet deep, as disappointed Dutch farmers had found. It couldn't hold much for long. And that thin layer rested atop thick, fractured bedrock of schist and Fordham gneiss. That bedrock later proved useful for carrying the weight of skyscrapers, but it made underground water supplies dangerously vulnerable to the excreta that was dumped on top of it. Fresh human waste that sank through the thin soil into the bedrock entered an underground highway of cracks and fissures, through which it could travel hundreds of yards.[21]

These geographical features made drinking-water supplies in the city especially vulnerable to contamination. And supplies were limited to begin with. The Hudson and East Rivers that surrounded the island were too salty to drink. Collecting rainwater proved treacherous. By the time the rain passed over residents' filthy roofs, collecting ashes and soot, "its appearance is nearly as dark as ink," one local noted, "and its smell any thing but agreeable."[22] (This paucity of drinking-water sources had been noted as a serious drawback to settling the island as early as 1664. The Dutch fort, its last governor, Peter Stuyvesant, complained, subsisted "without either well or cistern.") The island's sole source of easily accessible drinking water was the seventy-foot-deep Collect Pond, a small kettle pond that had been gouged out by a retreating glacier. But as the city's population expanded northward, noxious industries such as tanneries and slaughterhouses were pushed out to the Collect's shores. Soon the pond had become a "very sink and common sewer," as one resident complained in an open letter to the city published in the *New York Journal*. In 1791, the city purchased all claims to the pond, and health commissioners called for it to be drained entirely. Workers cut canals and ditches to drain the springs that fed the pond. In 1803, the city ordered the drained pond filled, paying New Yorkers 5¢ for every cartload of fill—that is, garbage—they dumped into it.[23]

After that, New Yorkers had to make do with groundwater—the water that had seeped down below the surface—which they tapped at public wells on the street corners. These were dangerously shallow. Today's standards would call for at least a fifty-foot casing and further drilling below that to reach uncontaminated groundwater, if there was any to be had. Manhattan's nineteenth-century wells were only thirty feet deep. One such well, nestled amid the privies and cesspools of the city's most notorious slum, Five Points, daily provided seven hundred thousand gallons of groundwater to one-third of the city's residents, via a system of wooden pipes built by the Manhattan Company.[24]

New Yorkers knew that their drinking water was tainted. As a letter writer to a local paper noted in 1830,

> I have no doubt that one cause of the numerous stomach affections so common in this city is the impure, I may say poisonous nature of the pernicious Manhattan water which thousands of us daily and constantly use. It is true that the unpalatableness of this abominable fluid prevents almost every person from using it as a beverage at the table, but you will know that all the cooking of a very large proportion of this community is done through the agency of this common nuisance. Our tea and coffee are made of it, our bread is mixed with it, and our meat and vegetables are boiled in it. Our linen happily escapes the contamination of its touch, "for no two things hold more antipathy" than soap and this vile water.[25]

"Can you bear to drink it on Sunday's in the Summer-time?" a local paper complained in 1796. "It is so bad before Monday morning as to be very sickly and nauseating; and the larger the city grows, the worse this evil will be."[26] A local doctor noted that the city's well water so commonly caused diarrhea that it could be considered a cure for constipation, a "virtue derived from the neighboring sinks [cesspools]" and to "certain saline properties, which render it peculiarly efficacious in certain complaints."

In 1831, scientists from the New York Academy of Sciences (then known as the Lyceum of Natural History) found that compared to fresh river water from upstate, which held just shy of 130 milligrams of organic and inorganic material in each gallon, the city's well waters were positively semisolid, rich with more than 8,000 milligrams of debris per gallon. The water, even a former director of the Manhattan Company had to admit in 1810, was rich with its users' "own evacuations, as well as that of their Horses, Cows, Dogs, Cats and other putrid liquids so plentifully dispensed."[27]

Of course, New Yorkers didn't know that their polluted waters could transmit deadly diseases. But they did know that it tasted bad, so they rarely drank water straight. They turned it into beer or they added liquor, like gin, or they boiled it to make tea or coffee. Besides making the water more palatable, these preparations destroyed the fecal microbes in it. They could even kill cholera vibrio. At 20 percent concentration, gin can kill cholera vibrio in an hour's time. Vibrio perish in hot drinks, too.[28]

Unfortunately, it wasn't just the groundwater below that was contaminated with fecal matter. It was also the surface waters: the waters that lapped at the shores, flooded over the island, and stagnated in puddles on New Yorkers' streets and in their cellars.

⊙

Ironically, the contamination of the city's surface waters began with the voluntary attempt, on the part of some of the more wealthy residents, to clean their privies.

This practice could have helped clean up the drinking-water supply, except that the most convenient place to dump the contents of their privies was in the rivers. Because locals found the practice smelly and unsightly, city regulations required that the dumping occur at night. (This is how human waste came to be called "night soil.") That regulation made the process even messier than it would have been otherwise. The odiferous loads were ferried to the piers over dark cobblestoned streets on rickety horse-drawn wagons, spilling their contents along the way. Once the wagons got to the piers, they sometimes dumped their loads directly on unseen vessels moored nearby. "Small boats which may happen to be within reach of the avalanche are either wholly or partially filled," the city inspector reported in 1842, "and instances are said to have occurred of their being carried to the bottom with their unnatural load."[29]

Worse than the mess on the streets was the waste that collected underwater in giant mountains. Under more amenable geographic conditions, the dumped privy waste would have been flushed out to sea by the tides and currents. But the waters around the island of Manhattan were stagnant and the land marshy. While the downstream currents of the Hudson and East Rivers pushed water southward away from the island, the Atlantic tides pushed it back toward the island. One result was that the city had to regularly dredge the slips in order to keep them navigable. Another was the continuous contamination of the river's waters, as the submerged piles of waste enriched them with nutrients ideal for the growth of bacteria. "Where the sun fell on it, it was literally effervescing," a local noted in 1839, the river's waters "actually sending up streams of large bubbles from the putrefying corruption at the bottom."[30]

Residents were regularly exposed to the contaminated waters in their rivers. There was little distinction on the island of Manhattan between the land and the waters around it. (Geographically, Manhattan and its surrounding islands were the temperate equivalent of the Sundarbans. Both consisted of low-lying archipelagos situated in the middle of estuaries.) Even before the city had been established, the low-lying, narrow island had been regularly inundated by the sea. The Lenape, the island's original inhabitants, who'd been displaced by the Dutch, had been able to paddle their canoes from one side straight to the other at high tide. In the winters, ice skaters could glide from today's City Hall to Greenwich Village and out to the Hudson.

But while the original inhabitants of the island could escape the floodwater, because it quickly drained back to the sea, nineteenth-century inhabitants couldn't. War and urban development had razed the high ground and blocked the streams and canals that might have carried floodwaters away. After the Revolutionary War, nearly half of the island's trees had been burned down, in forest fires and in the frenzy of rebuilding that occurred afterward. The island, wrote George Washington in 1781, was "totally stripped of Trees, & wood of every kind." The only thing left was "low bushes." The northern part of the island, which had been dotted with more than five hundred distinct hills (inspiration for the Lenape name for the island, "Mannahatta," or "Island of Many Hills"), was flattened. Bunker Hill, which had once risen behind the Collect Pond, had been leveled. Streams, canals, and ditches that might have drained the floodwaters were either choked with garbage or paved over.[31]

The worst flooding occurred in the many low-lying, sinking properties that had been "reclaimed" from the sea. The city had sold parcels of sea and pond called "water lots" to entrepreneurs, who displaced the water and constructed housing atop it. More than 130 acres of land around the coasts had been thus reclaimed, turning the once pointy tip of Manhattan into a rounded arc.[32] So had the site of the former Collect Pond.[33] But these reclaimed acres weren't stable like those underlain with Manhattan's bedrock. The lots had been filled with garbage and soil, which compressed and shifted under the weight of the buildings constructed atop them, like loose gravel under a heavy brick. As the fill compressed, the properties themselves sank.

And so for most residents there was no avoiding the floodwaters when they washed over the city's streets twice daily with the tides. The polluted water festered in puddles and collected in their cellars and yards. It gurgled through the streets and trickled into wells. Inevitably, with excreta-contaminated surface waters all around their homes, and excreta-contaminated groundwaters filling their wells, human waste found its way into New Yorkers' bodies. All a pathogen like *Vibrio cholerae* had to do was finagle a ride into town.

⊙

Drought descended upon Manhattan during the spring of 1832. The island's fragile groundwater resources shriveled, increasing the relative proportion of nutrient-rich filth to fresh water. The rivers grew brinier, the sun beating down triggering blooms of plankton and the tiny floating copepods that fed on them.

Cholera arrived that summer.

The first reported cases began in encounters with the waste-enriched, plankton-choked rivers. On June 25, on the east side of the island, a tailor named Fitzgerald, who lived on the first floor of a house on Cherry Street near the East River, boarded a ferry bound across the river to Brooklyn. He sickened with cholera and infected his two children and wife, all of whom perished. A few days later, two miles away, on the west side of the island, a man named O'Neil, in a drunken stupor, fell into the Hudson River. Cholera killed him, too. At the same time, cholera struck a fishing vessel moored in the East River, just south of Cherry Street, as well as at 15 James Slip and on Oliver Street, at the time both waterfront properties on the East River within a couple blocks of Cherry.[34]

Cholera vibrio from those early victims' guts quickly spread into the city's drinking-water supply. It was so hot that summer that some New Yorkers drank plain water by the glass despite its universally despised taste. Each glass could invisibly hold 200 million cholera vibrios.[35] And while special preparations could kill cholera vibrio, corrupt vendors and bartenders could undo those protections. Penny-pinching vendors commonly diluted the milk they sold with water, and in cheap dives, bartenders diluted the alcoholic drinks with water, too. When New Yorkers drank their hot teas and coffees with water-diluted milk, or consumed diluted cocktails, they could get a deadly dose of cholera.[36] (While a 20 percent gin cocktail killed cholera vibrio in an hour, a 15 percent gin cocktail wasn't strong enough for the job.)[37] Those who didn't drink straight water or diluted hot drinks and cocktails were exposed through various food items. Oysters, a dozen of which were cheaper than a single hot dog, were likely sources of cholera contamination, as were fruits and vegetables that had been splattered with contaminated water when the city's markets were washed down.[38] Contaminated water distributed by the Manhattan Company was widely shared. New Yorkers filled their buckets at the taps and hauled it to their tenement apartments to share with their neighbors. Grocers gave it away for free to passengers on ships and to their customers, to drum up business.[39]

Whole families perished from cholera. At one makeshift cholera hospital, a husband and wife, the wife's mother, the son, the servant, and an uncle all died within four days of each other. At the foot of Warren Street, beside the Hudson, a shocked passerby wrote of a "filthy and wretched" underground apartment, where he found a child "writhing in pain, next to a pile of bed clothes." Five of the residents of the apartment had already died and their bodies had been carted away, leaving just the child and his mother. When the child was asked where his mother was, he pointed to the pile of rags next to him, where surprised physicians found her dead body.[40]

People who lived on reclaimed lands were especially hard hit. The residents of the Five Points slum built atop the filled-in Collect Pond were mostly poor immigrants and African Americans. Many ended up in makeshift cholera hospitals. At one such hospital, more than half died.[41]

The disease staggered the city's physicians. One reported of his examination of a couple in the throes of cholera in their home. The bed and linens were "saturated with a clear odourless liquid," and while the woman screamed incessantly for water, the man beside her lay prostrate as the doctor reached out to take his pulse. "The touch of the skin, so unlike that which I had ever felt before, though I have attended the death bed of many, thrilled me with horror," he wrote. "I could not believe that I had laid my hand on a yet breathing body." The hands of the doomed duo were wrinkled, like a "person who had been washing," the doctor wrote, "or perhaps still more those of a person who had been dead for many days."[42]

They comforted themselves that the disease seemed to preferentially strike areas of "a low, dirty, and unhealthy character, consisting chiefly of confined narrow streets, and dirty dwellings, inhabited by the poorest and lowest part of the population," as one put it at the time. But insofar as that was true, it was only because most prosperous New Yorkers had fled the stricken city. If given the opportunity, cholera slayed the wealthy as efficiently as it did the poor, killing a member of the city council and the daughter of the fur magnate John Jacob Astor, at the time the richest man in the United States. Wealthy people lived on reclaimed lands, too. In the grand house at 26 Broad Street, which had once been an inlet of the East River, three women and four nurses and servants—"young, healthy, and temperate women," their physician reported—died within days of each other. So did a four-year-old child who'd been brought to the house for an overnight stay. Each victim, physicians noted at the time, had "slept or were habitually employed in the basement rooms." Clusters of cases similarly emerged among prosperous New Yorkers living at the foot of Duane Street, and on Vestry and Desbrosses Streets, in what is now TriBeCa, built on land reclaimed from the Hudson.[43]

Soon, cholera was taking more than a hundred lives every day, raging violently in one-quarter of the city "until it appeared to have exhausted all its subjects," as a mystified New York city physician put it, and then emerging "immediately in some other, and perhaps quite distant part."[44] "I am not able to walk to St. Mark's Place," one resident wrote to her daughters, whom she'd sent out of town while the cholera raged. "The Cholera is all we hear and think about . . . It continues its dreadful ravages with unabated vigor."[45] "There was scarcely a morning, for two months," noted another diarist, a shopkeeper who lived on Cortlandt Street, "that I did not meet on Broadway from three to six ambulances, with cholera patients on their way to the hospital."[46]

By mid-July, the city had ground to a halt, silent and still save for the sound of carts transporting corpses to the cemeteries and the drifting smoke from the burning of the clothes and bedding of the dead.[47] Stores had closed their doors and the city had canceled its usual Fourth of July festivities. "The disease is here in all its violence and will increase," former mayor Philip Hone wrote in his diary. "God grant that its ravages may be confined and its visit short!"[48]

◉

To those of us who've grown up with indoor-plumbing systems that capture every last drop of excreta in gleaming porcelain portals and flush it miles away, the waste management crisis that engulfed nineteenth-century New York seems a curiosity from another world. But it isn't. The sanitary revolution that replaced privies and latrines with flush toilets and running water has been selective and only partially implemented. And because pathogens that enjoy transmission opportunities in one corner of the world can easily spread elsewhere, in certain ways we are as threatened today by pathogens that spread by feculence as we were almost two hundred years ago.

While Western attitudes toward human excreta have radically changed since the days of poudrette, our attitudes toward animal excreta remain relatively cavalier. For example, in the United States, where dog ownership is common, many consider dog feces harmless. That's why many communities allow resident pets to defecate at will in streets, yards, and parks, and dog owners think nothing of walking for miles with thin plastic bags of dog feces swinging casually at their sides. A worker in the garden center at Home Depot confided to me that he planted his award-winning tomato plants in the stuff. According to one survey, 44 percent of dog owners make no attempt to collect or contain their dog's waste, explaining that dog poo acts as a fertilizer.[49]

As a result, dog waste deposited on sidewalks and yards sinks into soils, wafts into the air, and washes into waterways. About a third of the bacterial contamination in U.S. waterways originates in dog waste, and such contamination is more common in residential areas where dogs live than in commercial areas. (Scientists call this phenomenon the "Fido Hypothesis.")[50] It's in the air, too. One study of outdoor air pollution in Chicago, Cleveland, and Detroit found that during the winter when trees are leafless (and therefore not exuding bacteria into the air), the majority of the aerosolized bacteria comes from dog feces.[51]

But far from being a harmless source of fertilizer, dog feces is both an environmental contaminant (and is classified as such by the Environmental Protection Agency) and a source of pathogens that can infect people. Like human excreta, dog poo teems with pathogenic microbes, such as strains of E. coli, roundworms, and other parasites. One of the most common parasitic infections in Americans is the result of their exposure to dog feces. The dog roundworm Toxocara canis is common in dogs and, because of the ubiquity of dog feces, widespread in the environment. It can contaminate soil and water for years. Kids pick up the parasite when they play in contaminated soil and inadvertently put their hands in their mouths. Although it's rarely diagnosed, since there's no easily available, effective diagnostic test for it, one recent survey found that 14 percent of Americans over the age of six were infected. In humans, Toxocara canis has been linked to asthma and a range of neurological effects.[52] Dogs can also carry the tapeworm Echinococcus multilocularis, which in humans causes a disease that resembles liver cancer. The infection is a growing problem in Switzerland, Alaska, and parts of China.[53]

This casual disregard for the microbial potency of pet excreta extends to the waste of livestock. Parents who purchase diaper wipes in bulk to rid the last trace of human feces from their children's bodies don't balk when, on visits to the farm and the state fair, their kids are led over walkways as caked with excreta as the streets of nineteenth-century New York. And we allow the livestock we eat to live in conditions that would be considered medieval if people were subjected to them. The chicken coop, the pigpen, the rabbit hutch—all are piles of excreta with animals sleeping and living on top.

In the past, livestock excreta were put to good use as fertilizer, in small farms where both animals and crops were raised. That was possible because the amount of animal waste roughly matched the absorptive capacity of nearby cropland. Not anymore. Today, farms produce far more waste than croplands can absorb. Livestock populations have grown much larger. In the United States, the mean size of hog farms grew by more than 2,000 percent between 1959 and 2007; the mean size of farms where chickens are raised for meat grew by over 30,000 percent.[54]

As a result, livestock produce thirteen times more solid waste than the human population does in the United States.[55] To cope with the tens of millions of gallons of excreta they produce, farmers mix it with water and then pump it into untreated, multi-acre cesspools ("manure lagoons"). This wastewater is sprayed on crops, but since local croplands can't absorb it all, it leaches into the groundwater and runs off into the surface waters. It also creates a plume of odor and a fine mist of polluted water that blankets the drying laundry, cars, and homes of those living downwind of the farms.[56] "You don't plan birthday parties outside," one resident who lived near a manure lagoon said. "You no longer plan things. You plan around the odor and flies." (One woman quoted by *The New York Times* remembers swatting more than one thousand flies in her office in a single day, thanks to a nearby manure lagoon.) Although there are federal regulations aimed at preventing animal waste from entering surface waters, they're poorly enforced. During storms, the cesspools overflow, spilling their contents into local waterways. In 2013 in Wisconsin, more than one million gallons of manure spilled into the environment. In one of the worst spills, after a 1999 hurricane in North Carolina, 25 million gallons of manure poured into a local river, polluting 9 percent of local drinking-water wells with fecal coliform bacteria and killing millions of fish.[57]

This widespread fecal pollution creates novel transmission opportunities for a host of new pathogens. Shiga toxin–producing *E. coli*, or STEC, is one. About half of all cattle on American feedlots are infected with the microbe, which can survive in the environment for weeks or even longer if the weather is cool. In humans, STEC causes bloody diarrhea and life-threatening complications including hemolytic uremic syndrome, in which the kidneys fail. Up to 5 percent perish, and a third of those who survive suffer lifelong kidney problems.

Since the first reported outbreak in 1982, STEC has struck in fifty countries around the world. Despite ongoing efforts to contain it, each year, STEC will infect seventy thousand Americans. Those who live in countries where industrial cattle farming is common, such as in the United States, Canada, Britain, and Japan, are most at risk, and the closer you live to a cattle farm, the higher the risk.[58]

But the risk extends far beyond farm areas, as feces-contaminated products get shipped and consumed all over the world. In 2011, a batch of fenugreek seeds in Egypt caused an outbreak of disease three thousand miles away in Germany. That outbreak was noteworthy for two reasons. It showed the long reach of fecal-contaminated products and how they pose a risk to everyone along the global food chain. It also showed how pathogens exploit fecally contaminated environments not only for their transmission opportunities but also to become more virulent.

This latter enterprise has to do with how microbes exchange genetic material. Unlike creatures like us, who exchange genes "vertically," from parents to children, microbes can exchange genes laterally, by bumping up against each other. Scientists call it "horizontal gene transfer." Since it happens in places where microbes meet, microbe-rich, fecally contaminated environments provide a conducive setting.[59]

This has allowed many pathogens to become more virulent. Horizontal gene transfer is the process by which cholera turned into a pandemic killer, when a bacteriophage (a virus that infects bacteria) bumped up against the vibrio, endowing it with the ability to secrete toxin. Horizontal gene transfer created MRSA, when *Staphylococcus aureus* acquired the ability to secrete a toxin called Panton-Valentine leukocidin from another virus, along with genes that allow it to resist antibiotics from related bacterial species. It's the process by which the NDM-1 plasmid moves between bacterial species, endowing them with its powerful drug-resisting capacities.[60]

The pathogen that caused the 2011 outbreak in Germany had gained its virulence through two horizontal gene transfers. First, a bacteriophage infected a harmless strain of *E. coli*, endowing it with the genes to secrete the Shiga toxin, creating STEC. A second horizontal gene transfer endowed the pathogen with the ability to secrete even more toxins and to resist a broad range of antibiotics. The result was an especially virulent strain of STEC called O104:H4, which triggers life-threatening complications in twice as many victims as regular STEC.[61]

Sometime before 2011, the pathogen spread onto the fenugreek farm in Egypt, migrating deep into the seeds' interior, where it could hide from the disinfecting solutions that farmers use before planting.[62] Fifty different companies in Germany bought sixteen tons of the invisibly contaminated seeds and sold them to gardeners and farmers across the country to grow into sprouts. In the spring of 2011, when people in and around the city of Hamburg ate the sprouts sprinkled on salads and crudités, O104:H4 slid into their bodies.[63]

Scores started pouring into clinics and hospitals, confused and unable to speak clearly. Their "awareness becomes blurred, they have problems finding words, and they don't quite know where they are," said the Hamburg nephrologist Rolf Stahl. The gastroenteritis they suffered included bouts of bloody diarrhea. Children went into seizures and had to be put on dialysis. One woman's large intestine became gangrenous, and the left side of her colon had to be removed. She had muscle spasms that left her unable to speak. It was a "completely new clinical picture," Stahl said.

By the end of the outbreak, four thousand people across Europe—primarily in Germany but also a handful in France—had sickened. Close to fifty died. Some survivors developed severe neurological symptoms, including seizures, as a consequence of the infection.[64]

We haven't seen the last of this pathogen. After O104:H4 completed its violent passage in people's bodies, it exited the same way it had in the cows. And it did so for months after the outbreak subsided, as survivors continued to shed the pathogen in their stool, introducing a steady stream of intact pathogens back into the environment, where they will mix and match with the other microbes encountered there.[65]

⊙

As we grapple with the pathogens this new sanitary crisis imposes on us, we also face the pathogens churned out by the old sanitary crisis, which continues unabated in much of the world where poverty is rampant and governance weak. Fast-forward 178 years after New York City's first cholera epidemic, to the island of Hispaniola, and the nation of Haiti. The majority of the population used the same waste-management methods as nineteenth-century New Yorkers. As of 2006, only 19 percent of the Haitian population had access to toilets or latrines. "When our children have to take a poop, we put them on a little bowl," explained a resident of Cité Soleil, Haiti's biggest slum. "Once they are done, we throw it into an empty lot." Others used what are euphemistically called "flying toilets." They defecate "into plastic bags which are then tossed into a mound of garbage or a nearby canal," as watchdog journalists from the NGO Haiti Grassroots Watch wrote. And the excreta deposited into Haiti's streets and empty lots aren't easily moved. The flow of rainwater that might wash it into the sea is regularly blocked by garbage, such as plastic bags, Styrofoam containers, vegetable scraps, and cast-off shoes in various states of decomposition.[66]

People who live in the slums of South Asia are similarly exposed to human excreta. Nobody pays any mind when little boys, like the one I met at the Ekta Vihar slum in New Delhi, nonchalantly squat over the open gutters that run through the illegal settlement, even as a limber sari-clad woman and her three small children crouching in the dust on the gutter's banks eat their afternoon meal not twenty yards away. Of more than 5,000 towns in India, only 232 have sewer systems that carry away human excreta, and even those systems are partial at best. Everyone else must relieve themselves outside, in the open, like 2.6 billion other people around the world. Or they may use dry latrines of some kind or another, which are periodically emptied by India's 1.2 million "manual scavengers," who, like New York's nineteenth-century night scavengers, collect the excreta with their bare hands or a piece of tin. They scrape it into a basket and carry it to a designated dumping ground, such as a nearby body of water.[67] Whether collected by scavengers or carried off by sewer systems, the overwhelming majority of human excreta in the developing world ends up in the same streams, rivers, lakes, and seas that people use for domestic purposes, with microbial intensity fully intact.[68]

For the billions of people who lack adequate sanitation, this problem is a standing public-health catastrophe. Nearly 2 million die every year from diarrhea and scores of others from diseases such as intestinal worms, the helminthic infection schistosomiasis, and blindness-causing trachoma, which sanitary waste disposal systems could prevent. But it's not a problem only for them. It's a problem for everyone, because neglected environments contaminated with human filth provide a back channel for pathogens to amplify, spread, and hatch new pandemics that can affect all of us.

In Haiti, it was cholera.

Ten months after a magnitude 7.0 earthquake devastated Haiti in January 2010, a contingent of United Nations peacekeepers arrived direct from Kathmandu, Nepal, which was then in the grip of a cholera epidemic. They were stationed at a camp above the Artibonite River in the mountains north of Port-au-Prince. The facility, nominally a UN camp, had been designed and built by the Nepalese soldiers. Because there was no sewer system in Haiti, they built their own waste disposal system as well. Locals had long known it to be faulty: raw sewage from the camp poured into a stream that flowed into the river. The camp's neighbors could see it and they could smell it, and reporters later documented it, too.[69]

This wasn't the only time that aid workers had seen no option but to deposit human waste into Haiti's waterways. Earlier in 2010, the Red Cross and other aid agencies had dumped untreated waste from fifteen thousand chemical toilets into an unlined open pit, the size of four football fields, atop the Plaine Cul de Sac aquifer, which supplied the capital city of Port-au-Prince its scarce drinking water.[70]

While there's no evidence the contents of those chemical toilets contaminated Haiti's water supply with pathogens, the Nepalese soldiers' sewage dumping did. Within days of the soldiers' arrival, cholera vibrio had been introduced into the Artibonite River. The cholera-contaminated river water washed into the river's delta, where thousands of Haitian farmers grew rice. The farmers lived knee deep in the delta's brackish waters, diverting it into their irrigation canals and scooping it up to bathe in and drink. They didn't stand a chance. Nor did the rest of the Haitian population, which having never been exposed to cholera had no immunological defenses against it. Within a year, there were more cholera victims in Haiti than anywhere else in the world.[71]

The lack of adequate sanitation in New Delhi similarly allowed NDM-1 bacteria to make their way into local waters. A 2010 study found NDM-1-endowed bacteria in 4 percent of samples collected from drinking-water supplies and 51 of 171 samples collected from puddles in streets and alleys.[72] Whether local people have contracted NDM-1 bacteria from contaminated waters in India is unknown, but it's plausible. Evidence suggests it has happened elsewhere.[73]

⊙

In a broader context, the real problem lies elsewhere. Waste becomes an issue for management only when the volume of waste grows beyond the space available to dispose of it. That is to say, it's a direct result of the size and density of human and animal populations. Filth is just a symptom. The real problem is crowding.

FOUR

CROWDS

Were it not for urban growth during the mid-nineteenth century, the 1832 cholera epidemic in New York might have been the city's last.

The intensity of the epidemic led to its own collapse. By the end of the summer, cholera had rid the city of susceptible victims. Reported cases, which likely comprised just 1 to 30 percent of the total number of cases that occurred (since mild cases generally go unreported), totaled more than 5,800. Reported deaths reached nearly three thousand. Taking into account the extent of unreported cases, cholera had likely infiltrated the entire city. The only people left were survivors of the infection.[1] These survivors, as modern experiments have demonstrated, would have been immune to the pathogen. Cholera wouldn't have been able to cause another epidemic even if post-1832 New Yorkers drank cholera-contaminated water by the gallon.[2]

And so New York City had gone back to business as usual. "The stores are all open, footwalks lined with bales and Boxes & streets crowded with carts & porter cars," wrote John Pintard, a prominent merchant, in a letter in mid-August. "What a contrast to the middle of July when this Bazar of our dry-goods [on Pearl Street] had appeared as still & gloomy as the Valley of the Shadow of death . . . Now all life & bustle, smiling faces, clerks busy in making out Bills, porters in unpacking & repacking Boxes, joy & animation in every countenance."[3]

Cholera wasn't gone, however. The vibrio likely hid unnoticed in the coastal and surface waters around the city. It might even have caused a few easily ignored sporadic and isolated cases. Or it might have retreated into what's called a "viable but nonculturable" state, a kind of suspended animation in which its cells shrink and stop replicating, biding their time until conditions improve. (Pathogenic bacteria in cow's milk and bacteria in wastewater retreat into such states when faced with the assault of pasteurization and chlorination, respectively.) Either way, cholera's presence concealed, the memory of the epidemic of 1832 faded into oblivion.[4]

Meanwhile, fuel for a new epidemic accumulated.

During the generation that grew up between the cholera epidemic of 1832 and that of 1849, a novel experiment in crowded urban living unfolded. People across Europe and North America were drawn into burgeoning cities like iron fillings to magnets. Between 1800 and 1850, urban populations in France and Germany doubled. During roughly the same period, London's population tripled. Between 1830 and 1860, urban populations in the United States grew by more than 500 percent, three times faster than the national population.[5]

Many flocked to the cities for new manufacturing jobs, which offered better pay and more security than the farm work they abandoned. But the economic changes that industrialization unleashed led to large-scale population movements in other, less expected ways, too. The events that brought scores to New York City, for example, began when a steamship carried a peculiar batch of potatoes into Ireland in 1845.[6]

⊙

Millions of impoverished Irish tenant farmers depended upon the potato—what they called "God's gift from heaven"—for their sustenance. The average Irish laborer consumed upward of ten pounds of potatoes a day, roughly the amount of potatoes that the typical modern American might eat in a fortnight. He'd burn more potatoes for fuel. This wasn't the result of any inherent love for the potato, but because it is starchy, calorific, and easy to grow. As a result of discriminatory English policies, Irish tenant farmers had only scarce, marginal lands with which to feed their families. The potato was the only crop they could afford.[7]

But their overreliance on the tuber made them dangerously vulnerable to any pathogen that might prey upon it. One such pathogen arrived in 1845, in a batch of potatoes infected with a fungal pathogen named after the Greek for "plant destroyer." *Phythophthora infestans* originated in the Toluca Valley in Mexico. It had never been seen in Ireland before 1845, because in the days of slow-going sailing vessels, any potatoes infected with it would have turned to mush before hitting shore. But after steam travel shortened the journey, the infested potatoes were able to arrive intact. And once planted, the pathogen within them spread, invisibly broadcasting its deadly spores to nearby plants. Infested plants looked normal enough, but underground, their roots rotted. When farmers rustled in the soil to pull up the potatoes, their hands emerged from the ground covered in smelly goo. They unknowingly ensured the pathogen's return by discarding the ruined potatoes in heaps, from which the fungus reemerged every spring, ready to destroy the next year's crop.[8]

As the potato crop wasted, famine set in. One and a half million perished. Another 1.5 million fled the *Phythophthora*-devastated countryside, their travel overseas subsidized by their own landlords, who preferred to facilitate their departure rather than contribute more to famine relief efforts, as the Irish government demanded.[9]

Between 1847 and 1851, nearly 850,000 refugees from Ireland landed in New York City.[10] Only the most prosperous continued on into the interior of the country. The rest, unskilled laborers and former servants with nothing to eat and nowhere to live, "with money scarcely sufficient to pay passage for and find food for the voyage," as a local Irish paper noted in 1847, settled in their port of arrival: Manhattan, which would soon become one of the most crowded places on earth.[11]

There was no room to sprawl on the island. Nor was there any network of quick transit that could link distant parts of the city with its booming manufacturing and port areas. Locals and newcomers alike needed to live near their work, or at least the possibility of it. Crowds formed around these centers of economic activity like barnacles on piers.

☉

Many of the Irish famine refugees settled in the neighborhood named after the five-pointed intersection at its center and built atop the garbage-filled Collect Pond: Five Points.

Their arrival set off a housing boom. To accommodate the new-comers, property owners built haphazard additions atop the neighborhood's original two-and-a-half-story wooden buildings. They built new houses in the backyards, crowding two or even three houses on twenty-five-by-one-hundred-foot lots. They converted stables into apartments and rented out attics and basements with windowless bedrooms too low to stand up in.[12] When that didn't satisfy demand for housing in Five Points, property owners started tearing down the old wooden buildings to make way for tenements: four- and six-story brick buildings specifically designed to cram as many people as possible inside. The first tenement had been built at 65 Mott Street in Five Points in 1824, jutting out above the buildings around it "like a wart growing on top of a festering sore," as a local reporter put it. In the back lots, they built more tenements, of necessity just half the size of the front buildings. These "rear tenements" had no windows on the backs or sides, so the sole apertures looked out into the dim, privy-filled alleys between the buildings, festooned with lines of laundry. Some property owners even squeezed third tenements into their lots, or erected little shacks in the yards by the latrines.[13]

The economics of the housing boom contributed to the crowding. Property owners didn't live in the dark, cramped domiciles they'd built. (Tenements were taller and deeper than the wooden buildings they replaced, so they were naturally darker inside, and owners rarely installed then available gas lighting.)[14] Instead, they leased their tenement buildings to "sublandlords," who ran saloons or groceries on the ground floors of the buildings and rented out the apartments. They earned a good living doing so, charging prices so high they routinely reaped a nearly 300 percent profit. But the high rents intensified the crowding, as cash-strapped tenants were forced to take on their own paying boarders to help cover the rent. Nearly a third of Five Points' residents lived with boarders.[15] In a typical tenement apartment on Cedar Street, a 144-square-foot room housed five families, sharing two beds among them.[16]

The most wretched of Five Points' domiciles lay underground. More than eleven hundred people in the Sixth Ward that comprised Five Points lived in basements, including in cellar flophouses, where a bunk—specifically, a piece of canvas stretched between wooden rails—could be rented for 37.5¢ per week. New York's doctors claimed they could spot basement dwellers by a glance at their pale faces and a sniff of their musty odor. It permeated "every article of dress, the woolens more particularly, as well as the hair and skin," as one put it.

In the past, the poor had lived dispersed along the periphery of towns and villages. Five Points reversed that pattern, drawing them all together. Sex workers found the centrally located slum convenient for accessing clients. Poor people across the city found shelter in its broken-down buildings. With an average per capita income that was the lowest in the city, gangs, crime, and prostitution flourished. Slums like Five Points became what the anthropologist Wendy Orent calls "disease factories." They could take a pathogen and whip it into an epidemic as ably as an engine combusted fuel into motion. And in Manhattan, the Five Points disease factory blazed not in some far-off, isolated locale but in the heart of the city.

City officials were only dimly aware of the health hazard that Five Points posed to the rest of the city. At one point, they considered razing parts of the neighborhood to build a prison. But they didn't, out of fear that the unsavory locale would lead to outbreaks of disease among the inmates. Most outsiders treated Five Points as a detached spectacle, threatening to them mostly in terms of their moral sensibilities. Journalists and writers toured the neighborhood, in a trendy practice they called "slumming," to pronounce their condemnation. (An echo of their disgust can be heard today, in the 2002 film *Gangs of New York*, directed by Martin Scorsese and based on a 1927 book about Five Points.)[17]

And so while commentators issued regular bulletins broadcasting their moral distaste (Charles Dickens called the slum "hideous" and "loathsome"), the crowds that would ignite a city-enveloping epidemic continued to grow. By 1850, in the slums of New York City, nearly two hundred thousand people crammed into each square mile. That's nearly six times more crowded than modern-day Manhattan or central Tokyo, and over a thousand times more crowded than any group of humans had ever lived before.[18]

⊙

After a seventeen-year absence, cholera returned to New York City in 1849 with a vengeance. The epidemic began, like many epidemics, with small, barely noticeable outbreaks on the outskirts of the city. In the winter of 1849, a packet ship called the *New York* arrived in New York harbor from Le Havre. Seven of the passengers had died of cholera during the journey. A city health officer hustled the ship's three-hundred-odd passengers into a customs warehouse, turning it into a makeshift quarantine hospital. Over the following weeks, cholera sickened 60 and killed over 30 at the warehouse-cum-hospital. Unbeknownst to the rest of the city, 150 others scaled the warehouse walls, boarded small boats, and escaped into the metropolis.

In January 1849, cholera broke out in the city's immigrant boarding-houses, possibly as a consequence of the infected escapees' arrival. A lull of a few months during the winter followed. Then in May, cholera infiltrated Five Points. In rooms without running water where multiple families prepared food, ate, and slept together, the vibrio easily passed from person to person. Cholera stool clung to hands and splashed on shared bedding and clothes, which residents sold to ragpickers and washed at public taps. Like a hurricane hovering over warm waters, the epidemic gained strength.

Once the vibrio entered the groundwater, cholera exploded across the city. (Although piped water from unpolluted upstate sources had become available in 1842, two-thirds of the city continued to rely on the shallow public wells on the street corners.)[19] The health department closed four public schools to use as cholera hospitals, casting schoolchildren out into the cholera-plagued streets. Dead bodies lay untended for hours and sometimes days before being picked up and taken to the potter's field on Randalls Island, where they were deposited the same way they were in Paris in 1832, in wide, shallow trenches, one atop the other.

By the summer, President Zachary Taylor could do little more than call for a day of "national prayer, fasting and humiliation" to tame the city's raging epidemic. Ultimately, more than five thousand would die.[20]

⊙

By all rights, the urban experiment that began in the nineteenth century should have failed. By the middle of the century, writes the historian Michael Haines, big American cities had become "virtual charnel houses," their primary demographic characteristic being high mortality. Deaths outnumbered births. Despite the greater availability of food and paid work, children under the age of five who lived in cities died at nearly twice the rate as those living in the countryside. In 1830, a ten-year-old living in a small New England town could expect to see his or her fiftieth birthday. That same child, living in New York City, would be dead before the age of thirty-six. If you plotted population density and early childhood mortality on a graph for the years between 1851 and 1860 in England and Wales, you'd see a straight line heading upward.[21]

Even those who survived suffered the price of urban living. Their poor health stunted their growth: the average height of West Point cadets born between 1820 and 1860, as the nation became more urbanized, declined by a half inch. The shortest recruits came from the most densely crowded cities. The same process of deterioration unfolded in Manchester, Glasgow, Liverpool, London, and everywhere else that crowded urban living took root.[22]

Industrial cities survived, like dying patients on life support, because new blood in the form of immigrants kept pouring in to replenish their diminished, dying masses. In the years after the 1849 cholera epidemic in New York, immigrants continued to stream into the city, at the rate of nearly twenty-three thousand every month. They were more than enough to replace the parade of corpses flowing out of the city.[23]

Meanwhile, new regulations on housing slowly eased the deadliness of the city. The crusading journalist and photographer Jacob Riis used the new technology of flash photography to capture images of the dark corners of the tenement world for an aghast public. His 1889 book *How the Other Half Lives* helped ignite a movement for tenement reform in New York City. One of the first reforms, the Tenement House Act of 1901, required that city buildings provide exterior windows, ventilation, indoor toilets, and fire protections.

Five Points, a neighborhood premised upon crowding, didn't survive the era of housing reform. Much of it was simply demolished. A sliver of the old neighborhood became what is today Chinatown. Another, the site of the old Collect Pond, became a small paved park fenced with chain link, surrounded by imposing government buildings: Superior Court, City Hall, and the clinics of the Department of Health of the City of New York, among others. Passersby would never suspect that a riotous neighborhood of any kind once existed there.

The last vestiges of the slum were lost in the September 11, 2001, terror attacks, when 6 World Trade Center collapsed. The sole collection of artifacts from Five Points—eight hundred thousand fragments of porcelain, bone china, tea sets, tobacco pipes, cisterns, and privies that archaeologists had collected before the five-way intersection was demolished to make way for a courthouse in the early 1990s—had been stored in the basement.[24]

⊙

Thanks to the housing revolution, even the most crowded cities can be healthful places to live. In general, people who live in cities today live longer than those who live in rural areas. Only a few health burdens remain—higher rates of obesity and more exposure to pollution, for example.[25]

And yet, as washed clean of their past as cities like New York may appear, the housing revolution they enjoyed, like the sanitary revolution, has been partial and selective. It hasn't penetrated many of the poorer countries of the world, and its insights haven't been applied to our livestock. In India, due in part to poverty and in part to a lack of governance, housing regulations are as sparse and poorly enforced as in nineteenth-century New York.

In Mumbai, the densest streets in slums such as Dharavi hold 1.4 million people in each square mile, more than seven times the concentration of humans packed into nineteenth-century Five Points.[26] Rural migrants live on the street in shacks made of scrap metal and tarps. They cluster around the entrances of the city's middle-class apartment blocks, like the one my cousin lived in. I remember one morning, some years ago, while sitting by the grated window of his flat drinking tea, hearing a loud whoosh followed by a cloud of dust and some commotion from the street. The narrow cement terrace on the floor above us had sheared off the building and plummeted to the alley below, reassembling itself into a mound of steaming rubble. My aunt and cousins marveled quietly for a few minutes at the destruction, with about the same amount of interest and lack of alarm as might be provoked by a crow that had grabbed someone's toast.

Such scenes of urban decrepitude will become more common in the future, for the process of urbanization that began in the industrial era is accelerating. Back then, urbanization was rapid, but it was still exclusive: globally, more people lived outside cities than inside them. By 2030, experts estimate, that will change. The majority of humankind will live in large cities.[27] Only a handful of these large metropolises will be as healthful and well regulated as the cities of Europe and North America. Many will be more like Mumbai. Two billion of us will live in slums like Dharavi.[28] Our booming livestock population, which is larger today than the cumulative population of the last ten thousand years of domestication until 1960, live in the animal equivalent of slums, too. More than half of the world's pigs and chickens are raised on factory farms, and more than 40 percent of the world's beef is produced on feedlots, where animals are crowded together by the millions.[29]

The growth of slums is one reason why the 2014 Ebola epidemic was so deadly and long-lasting. Before 2014, Ebola outbreaks had never occurred in towns larger than a few hundred thousand. Four hundred thousand people lived in Kikwit, the Democratic Republic of Congo town that experienced an Ebola outbreak in 1995. Just over one hundred thousand lived in Gulu, Uganda, in 2000, when Ebola emerged there.[30] Since these locales were relatively small and remote, experts widely considered the virus, as the title of a 2011 scientific paper put it, a "minor public health threat" in Africa.[31]

But then the virus spread into West Africa, where it affected a markedly different demographic landscape. Ebola struck three capital cities, with a combined population of nearly 3 million: Conakry, the capital of Guinea, on the west coast of Africa; Freetown, the capital of Sierra Leone, 165 miles south of Conakry; and Monrovia, the capital of Liberia, 225 miles south of Freetown. These are not cities full of spacious high-rise apartments equipped with Wi-Fi and all the latest mod cons. They're overcrowded, haphazardly developed, and chaotic, as scores of news consumers finally learned when lurid photos of West African slums splashed across websites and newspapers during the outbreak.[32]

Crowds provide pathogens such as Ebola and others with at least three advantages. For one, they achieve a sharp uptick in their rates of transmission. When Ebola lurched out of Guéckédou and into the crowded capitals of Guinea and Liberia, its transmission rate spiked.[33] (The same thing had happened to variola, the virus that causes smallpox, when it emerged in urban centers, and could happen to its cousin monkeypox if, as the ecologist James Lloyd-Smith speculates, it spreads into a city like Kinshasa in infected meat or in the bodies of infected people.)[34]

For another, pathogens can burn through these larger populations for much longer. Each of the twenty-one outbreaks of Ebola that preceded the 2014 epidemic had been contained within a few months. Ten months after Ebola struck West Africa and its bristling cities, not only was the epidemic not under control, it was still growing exponentially. More than three thousand were dead; over six thousand had sickened. "We have never had this kind of experience with Ebola before," said David Nabarro, who was coordinating the United Nations' response to the epidemic. The nature of the urban landscape spelled the difference. "When it gets into the cities," he said, "then it takes on another dimension."[35]

But the most transformative effect of crowds lies in the way they allow pathogens to become more deadly. This has to do with the peculiar evolutionary advantages that pathogens that infest crowds enjoy. Under most circumstances, virulence is detrimental to a pathogen's ability to spread. Consider pathogens that spread when people breathe on each other, like influenza, or when they touch each other, like cholera or Ebola. Successful transmission depends on social contact between infected and noninfected people. Uninfected people must inhale the breath of the infected or touch their bodily fluids. If they don't, the pathogen is stuck. It can't spread.

This reliance on social contact makes virulence problematic for such pathogens. If they are highly virulent, their victims will sicken and perhaps even die. Infected people will end up alone in bed or isolated in hospital wards rather than at work shaking people's hands or on the train breathing on other passengers. When infected victims die, their bodies will be abandoned, burned, or buried—possibly before the pathogens lurking inside can spread to anyone else. This is a serious disadvantage. And it's why highly virulent strains are more likely than less virulent ones to die out. Virulence is evolutionarily constrained.

But certain human behaviors lift these brakes on virulence, allowing even the most deadly strains to flourish. One example is burial rituals that require the bereaved relatives to handle the corpses of their loved ones. A funeral tradition among the Acholi people of Uganda, for example, calls for corpses to be bathed by relatives and for their faces to be ritually touched by mourners. Similar rituals, which likely played an important role in the 2014 Ebola epidemic in West Africa, free pathogens from the debilities of virulence. Even pathogens that promptly kill their victims, like Ebola does, can spread into new victims, because social contact continues even when infected people are dead.[36]

Crowds of people in slums and animals in factory farms do the same thing. In crowds, social contacts that spread pathogens continue even when victims are sick and dying. The sickbed is in the living room or the kitchen where friends and relatives have easy access to the ill. The hospital wards are full and beds crowded with several patients each, worried relatives hovering at their sides. Sick animals are crammed into cages with healthy animals. Under such conditions, pathogens that evolve to become more virulent suffer none of the debilities that virulence would normally exact. They can spread regardless of how ill they make their victims.[37]

They can be as virulent, in other words, as the most dangerous pathogens in the world: those that don't rely on social contact to spread. These pathogens are either stable in the environment or are carried by vectors. They include killers like cholera; *Mycobacterium tuberculosis*, which causes tuberculosis; and variola, which causes smallpox. Virulence doesn't handicap their ability to spread, because they can spread from their dead victims by persisting in the environment until another live victim picks them up. The same is true for vectorborne pathogens like *Plasmodium falciparum*, which causes malaria. So long as mosquitoes keep biting, the pathogens continue to spread, no matter how sick their victims become. (On the contrary, virulence might even improve their transmissibility, since sicker, bedridden victims may be more likely to get bitten by mosquitoes compared to less ill ones.)[38]

Pathogens that spread through social contact are usually destined to be relatively mild. Crowds allow even these pathogens to become killers.

⊙

The way crowds make pathogens more virulent can be clearly seen in the case of influenza. In recent years, by providing the virus with large crowds of animals and people to infect, we've created a host of new, more virulent strains of the virus.

Influenza viruses originate in wild waterfowl and have long spilled over and adapted to other species, including humans. There are three types. Type B and type C influenza viruses are human-adapted pathogens, which cause mild seasonal flu. Type A are those viruses that have remained in their original reservoirs: ducks, geese, swans, gulls, terns, and waders.[39]

Occasionally, type A viruses spill over into domesticated poultry flocks. It's especially common in southern China, where traditional farming practices allow domesticated ducks and wild waterfowl to easily mingle, providing ample opportunities for influenza to spread into the domesticated flocks. But the domesticated birds, unlike the wild ones, have no immunity to flu viruses. The pathogen replicates with abandon in their bodies, evolving new, more deadly strains called "highly pathogenic avian influenza" or HPAI.[40] The process is so reliable that scientists can reproduce it in the lab, creating more deadly strains of avian influenza simply by passing the virus repeatedly through chickens.[41]

One important check on the spread of highly pathogenic avian influenza is the size of the domesticated bird flocks they infect. Infected chickens shed the virus in their feces for just a handful of days before the virus kills them. Without masses of susceptible poultry around, transmission dies out on its own within a couple weeks, as mathematical models show. In low-density poultry-raising regions, the basic reproductive number of these deadly viruses is less than 1.[42] That's why, up until 2000, scientists considered avian influenza the way they considered Ebola before it emerged in densely crowded West Africa in 2014, as an "infection of minor importance."[43]

But then the number and size of poultry farms in China started to grow. By 2009, nearly 70 percent of China's "broilers"—domesticated chickens raised for meat, as opposed to those raised for the eggs they lay—were reared on farms with more than two thousand other birds. Even larger farms became more common, too: between 2007 and 2009, the number of large farms with more than one million birds grew by nearly 60 percent.[44] The pace of international trade in poultry picked up. Twenty times more chickens were shipped across national borders in 2008 than in 1970.[45]

With bigger flocks and more poultry on the move, contact increased between poultry and wild bird territory and flyways, precipitating more frequent spillovers of flu from the wild birds to the domesticated ones. These spillovers in turn led to more frequent emergence of highly pathogenic avian influenza viruses, which were able to cause longer and bigger outbreaks in the larger flocks they infected. As a result of these changes, the deadly viruses crossed the threshold to become self-sustaining epidemic pathogens in poultry. According to mathematical models, the basic reproductive number of avian influenza viruses, in places where poultry farming is intense, is more than 10.[46]

As the virus's basic reproductive number increased, so did the frequency and scale of the outbreaks it caused. Between 1959 and 1992, outbreaks of deadly avian influenza had occurred roughly every three years. Most affected fewer than five hundred thousand birds. Between 1993 and 2002, outbreaks occurred once every year, and between 2002 and 2006, they occurred every ten months. About half of these outbreaks affected millions of birds at a time.[47]

For years, the growing viral menace posed by our supersized poultry farms escaped public notice, mostly because highly pathogenic avian influenza viruses were restricted to birds. They didn't infect humans. Then, in 1996, a wild bird flu crossed over into domesticated geese on a small farm in Guangdong province, one of the largest poultry-producing regions in China.[48] The virus, dubbed "H5N1," evolved two never-before-seen capacities. Unlike other avian influenzas, which were rarely found in wild birds, this virus struck a wide variety of wild species, including migratory ones. It also could infect humans.[49]

☉

People were exposed to H5N1 through close contact with infected birds. Normal flu symptoms would turn into severe pneumonia, and for some, organ failure. More than half—59 percent—of those infected perished.[50] And the virus spread. The international poultry trade carried H5N1 into poultry flocks in at least eight countries, including Thailand, Indonesia, Malaysia, and Cambodia.[51] Migratory birds ferried H5N1 into the Middle East and Europe.[52] As of this writing, North America has been spared, because few bird species migrate between North America and the H5N1-afflicted parts of the Old World. But that could change. The virus has intermittently been found in migratory birds in Siberia, who mingle with the ducks, geese, and swans that migrate across the Bering Sea into North America. If the latter birds get infected, North America could face H5N1 as well.[53]

Of all the new pathogens emerging today, novel influenza viruses like H5N1 are the ones that keep the most virologists up at night. If H5N1 or any other novel avian influenza evolved to transmit effectively between humans, the death toll would be swift and substantial. Even with low mortality rates, seasonal flu viruses carry off huge numbers of victims, simply because they are so good at spreading widely among us. Every year seasonal influenza kills up to half a million people around the world. That's the toll of flu viruses that have already adapted to us, and we to them. A novel influenza virus that could spread as well as the seasonal flu with an even marginally higher mortality rate could level millions.

For now, H5N1 is a zoonotic pathogen. It can't easily spread from one person to another, which is why of the tens of thousands of people who'd likely been exposed to the virus by the summer of 2014, only 667 cases in humans had been reported.[54] But as it evolves, H5N1's transmissibility in humans could improve. So far, the virus has evolved into at least ten distinct lineages, or "clades," all with varying abilities and proclivities.[55] Some have already mutated in ways that scientists think could boost the effectiveness of H5N1's transmission in humans. A distinct clade in Egypt, for example, appears better than any other at binding with human cells. That is probably why, between 2009 and 2013, more than half of those who sickened with H5N1 came from Egypt.[56]

And the virus continues to evolve. If an avian influenza like H5N1 does graduate from a zoonotic pathogen into a human-adapted one, the necessary adaptations will likely occur inside the bodies of people who work intimately with infected birds. I flew to Guangzhou to get a sense of the openings and obstacles the virus would encounter in their bodies. On the plane, it was apparent that flu season had descended upon us. Indeed, a couple weeks earlier, H5N1 had killed a thirty-nine-year-old bus driver in Guangdong province, setting off a massive slaughter of poultry in Hong Kong, many of which had been reared in southern Chinese provinces. The entire aircraft seemed to be coughing, creating an orchestra of high coughs, low coughs, long coughs, short coughs, wet coughs, and dry coughs. The habit of coughing into elbows had not caught on among my fellow passengers. People leaned down and let it all out. The tall young man sitting next to me studied the newspaper avidly during the flight. After we landed, he held a vomit bag to his face, coughed up a wad of mucus, spit it out, and stuffed the bag back into the seat pocket.[57]

When I arrived at the Jiangcun poultry market in Guangzhou, an open-air wholesale operation where thousands of chickens, ducks, and geese are reared in mesh-wire enclosures, there was no warning about the ongoing H5N1 outbreak, or the death of the bus driver. The birds seemed healthy enough, although according to a 2006 study, one out of every hundred was in fact infected with H5N1.[58] If the workers at Jiangcun were aware of this factoid, they didn't show it. In the United States, poultry workers handling birds infected with highly pathogenic avian influenza viruses wear the same elaborate protective gear that clinicians wear to fight Ebola.[59] The workers at Jiangcun were bare-handed and unmasked, wearing simple rubber boots and aprons. The middle-aged couples who managed each enclosure—about the size of the typical monkey cage at a zoo—used long metal hooks to grab the birds by the neck, unceremoniously stuffing them into plastic bins, which were then loaded onto trucks. The dead ones they simply stuffed into a tightly lidded blue plastic barrel. One of these stood beside every enclosure.

Their nonchalance stemmed, I figured, from the learned indifference of proximity. The workers lived bathed in the viral-contaminated excreta of the birds they tended. They spent their days in the enclosures, grew vegetables in the trash-strewn empty lot behind the caged birds, and retired each night to the low, stained cement-block flats just hundreds of yards away on the same property. A cloud of dust—bits of feathers, sand, and desiccated bird shit—had settled over everything: the lines of damp gray laundry strung over the concrete paths and the sagging cardboard boxes of packaged noodles and biscuits in the little shops across from the enclosures. The windows in the workers' flats were opaque with it.

The ragpickers who tended the market were similarly exposed. They lived in hand-built shacks covered with tarps just outside the market's fence. I saw them ambling through the market, shovels slung over their shoulders, collecting wet mush from the bird enclosures into eight-foot-tall piles. Piles of the stuff lay next to their shacks, too.

Any virus in bird excreta at Jiangcun poultry market enjoyed bountiful, unfettered access to the workers there. It could enter the human body as freely as a stream flows into the sea.

⊙

Precisely how H5N1 gained the ability to infect humans is not clear. Some experts speculate that another crowd of livestock—pigs—may have played a role. One of the biological barriers to the spread of bird influenzas in humans is the fact that bird-adapted viruses bind to sialic acids in birds that are not present in humans. Theoretically, H5N1 or some other bird-adapted virus could spontaneously mutate in such a way that allows it to bind to human sialic acids. But there's another, much faster way for novel flu viruses to acquire that ability, through what's called reassortment. This is when a virus acquires a chunk of new genes from another virus, and with the newly acquired genes, all of the capabilities those new genes confer. An avian influenza virus could reassort with one that was already good at infecting people, for example one of the many influenza viruses already adapted to humans, such as the relatively mild ones that cause seasonal flu. Then the novel avian influenza virus could acquire the ability to transmit efficiently in humans, too.

Reassortment of this kind could happen only in cells coinfected with both viruses at the same time. But since human influenzas bind to human sialic acids, and avian influenzas to avian sialic acids, people aren't easily infected with bird influenzas, and birds aren't easily infected with human influenzas. Thus, even with the massive flocks of poultry moving across international borders and thousands of people exposed to bird excreta across southern China and elsewhere, opportunities for human and avian flu viruses to directly exchange genes are slim.

This is where the pigs come in. Pigs have both humanlike sialic acids on the surfaces of their cells as well as avian-like ones. That means that both kinds of virus can bind to their cells. (This is also true of quails, but given the small scale of quail farming does not seem to play much of a role in the epidemiology of influenza.) Pigs living in proximity to both humans and poultry flocks or wild waterfowl could be the mysterious missing link between bird viruses and human influenza pandemics. Virologists call them the perfect "mixing vessel" for novel influenza strains.[60]

And just as poultry flocks have grown in China, so have the size of pig farms, increasing the probability that an avian influenza could spill over into pigs.[61] Up until 1985, 95 percent of China's pigs were raised in rural households that reared only one or two pigs a year.[62] By 2007, more than 70 percent of China's pigs were reared in farms where hundreds of pigs are crowded together. By 2010, China was the world's largest producer of pork, raising 660 million pigs, half of all the pigs raised in the world, and more than five times the production of countries such as the United States.[63]

A few days after visiting Jiangcun, I arrived in Laocun, Gongming, an illegal pig-farming colony in a sparsely populated industrial area about an hour outside Shenzhen. About one thousand pig farmers live there, along with thousands of pigs, in long, low shacks made of scrap metal and bamboo, paying rent for the government-owned land, illicitly, to the son of a Communist Party member. Along the lanes, groups of hens and a few sickly pigs turned out from the herds roamed freely. The pig farmers wore tall boots, preparing the refuse that they collect from restaurants and elsewhere into viscous stews, steaming in tubs, to feed to the pigs.

We drove around slowly, trying not to arouse attention, until we happened upon a ruddy-cheeked couple, who beckoned us into their shack. It was here that I saw how easily pig, bird, and human influenza viruses could mingle. The couple had divided the dirt-floored structure, rich with the odor of the pigs, into several dark enclosures sparsely furnished with old mattresses, some sticks, and used plastic bags. In one, a young woman hunched beside a smoky fire ringed by stones, a bucket of water by her side, vegetables floating atop. The family's laundry hung alongside one low wall. A pair of improbably bright red sneakers was stashed on a high wooden shelf.

About ten yards away, under the same roof, hundreds of pigs shuffled and snorted. There were at least three hundred pigs there, packed into long, narrow pens with a two-foot mud track in between them. Each weighed two hundred pounds or more, and most were huddled on top of one another, fast asleep, bodies caked with dung and the remains of their meals. Their heads, with their heavy chins and long, floppy ears, seemed immense atop their squat bodies.

As we made our way through the pig shack, we could see just beyond it a wide, shallow pond. The farmers told us they used this as a manure lagoon in which to store pig waste and raise fish. It was as still and shallow as the ornamental duck ponds that American developers build outside suburban malls. It undoubtedly attracted passing water-fowl. Their viral-contaminated droppings, as they flew overhead, could easily land in one of the many pig troughs that dotted the colony. In the summer, the farmers removed the scrap metal roofs on their shacks to let in the breeze, exposing pigs and their feed to the open air.

Seeing that lagoon, I could easily imagine a pig at Laocun coming down with both an avian influenza and a human one. In that animal's body, the next pandemic flu virus could hatch.

⊙

Whether or not H5N1 evolves into a human pathogen or fades away into oblivion, the risk of novel influenza viruses remains, as the growing crowds of people, birds, and pigs continue to hatch new strains with pandemic potential. Over the course of writing this book, at least two novel influenza viruses are known to have emerged, both with newly evolved capacities to infect humans.

A variant of H3N2, a virus that normally infects pigs, started spreading into people in the United States in the summer of 2012. (Scientists categorize influenza viruses by the type of proteins on their surfaces. Each has one of sixteen subtypes of the protein hemagglutinin (H) and one of nine subtypes of the enzyme neuraminidase (N) on its surface.)[64] People were infected with the pig virus at state fairs, where hundreds of pigs from across the state were gathered together in pig barns. The crowds of pigs created what the virologist Michael Osterholm called "airborne clouds of virus" inside the barns, which were readily inhaled by local people.[65]

I saw these firsthand at my local state fair in Maryland. People passed in and out of the pig barns, and the airborne clouds of virus that wafted inside them, as nonchalantly as the poultry workers and pig farmers lived amid infected bird excreta in south China. Onlookers walked freely among the small dusty pens in which the pigs slept, holding plastic cups of beer and leaning over to fondle the animals. Giant fans blew the barn's hot, dusty air around, ruffling visitors' hair. "Look at the piggy!" I overheard a teenage girl say to her friends, her shoes caked with pig muck. "He's so cute! Look how beefy he is!" Some of the pig handlers had annexed some of the pigs' pens, in lieu of a pricey hotel room. In one pen, a couple with two small girls had set up folding chairs and were munching on potato chips; in another, a pile of saggy mattresses sat atop bales of hay, covered with blankets and pillows. Whoever slept there inhaled the virus-laden air all night long.

Between 2011 and 2012, H3N2v from pigs managed to infect 321 people.[66] It's not a huge number of people, but for a pig virus that had never previously infected humans, H3N2v's progress into Homo sapiens was "unprecedented," Osterholm said. The virus had surmounted the species barrier and with repeated exposures could produce mutants with the ability to replicate inside humans. "We're tempting fate," Osterholm said.[67]

Another new influenza strain started infecting humans in February 2013, in eastern China. This virus, H7N9, was discovered in three patients hospitalized for severe pneumonia. Phylogenetic analyses suggested that the new virus was the product of multiple reassortment events involving viruses from ducks, chickens, and wild migratory birds that had occurred over the previous year somewhere near Shanghai. The changeling virus then likely amplified in poultry flocks.

Virologists worried about the spread of H7N9 because there'd been no glimmer of disease in the poultry themselves. Human cases of H5N1 were accompanied by concomitant outbreaks in domesticated flocks, providing a crude sort of advance warning. There'd been no such warning for H7N9. Because infected birds didn't get sick, the infections in people appeared as if out of nowhere. The virus seemed to have the capacity to spread silently, without manifesting in disease, in people, too. One study found that more than 6 percent of poultry workers harbored antibodies to H7N9, despite having no history of suffering an infection.

In the fall, a second wave of human infections began, this time over a much wider region, including southern China as well as eastern China. Since most of the people who caught the virus had been exposed to live poultry, it's likely that birds were the stealthy culprits behind its spread. By February 2015, H7N9 had infected more than six hundred people.

Like H5N1 and H3N2, it has not achieved the easy transmissibility among people that would make it a candidate for a pandemic—yet. Whether one of these novel viruses, or any of the other ones that continue to emerge from our supersize poultry and pig farms, will end up with the right genetic combination to do that remains to be seen.[68]

⊙

The most nightmarish flu pandemic in modern times struck in 1918. The pandemic virus—H1N1—had amplified and grown virulent under the unusually crowded conditions of trench warfare during World War I. It caused more than 40 million deaths around the globe, mostly due to bacterial pneumonia, a complication of the viral infection (which would be treatable today, unless caused by a resistant strain).

H1N1 sank out of sight after that. It seemed as if the virus disappeared. But it hadn't. It had retreated into some repository somewhere, just as cholera had in New York City in the fall of 1832. And just as cholera had, it remained quiescent until a sufficiently large crowd of susceptible humans formed, allowing it to strike out again. That happened nearly a century later, in 2009, precipitating the less deadly but still potent "swine flu" pandemic of that year.

The virus's century-long hideout, the virologist Malik Peiris told me when I met him in Hong Kong, was the bodies of pigs.

I thought of this as we left the pig shack in Laocun, shuffling single file behind the farmers down the dark path between the pig enclosures toward the door. One pig in the far end of the enclosure had roused and started climbing over the others toward us. Propped up at the pen's gate, he reared up suddenly at shoulder height, tilting his large head to turn one wide, almond-shaped pale-green eye into mine. It was as if he had something urgent to say, and I held his gaze for a moment. But all he did was emit a terrible sound, like the low roar of a howler monkey. I looked away and shuffled along behind the others, heart pounding. In the subtly shifted air our movements left behind, the pathogens brewed in his crowd mingled with those brewed in mine.

FIVE

CORRUPTION

The pathogen that can spill over, spread, and cause disease is a dangerous creature to be sure, but it's actually only halfway on the multistage journey toward pandemicity.

The fate of the second other half of its journey is determined by how societies respond. It's true that sometimes pathogens crash like a tidal wave, descending too quickly or harshly or cryptically for societies to understand what to do before it's too late. But in many cases, collective defenses of even the most crude sort—isolating the sick and warning each other of a disease's spread, say—can act like an underwater barrier, breaking the wave of death and destruction.

That levels the contest between pathogens and humans. Human cooperation is, biologically speaking, a formidable thing. Most mammals cooperate with each other only when they're related by blood. Not us. We cooperate more frequently, more intensively, and on a larger scale than any other species on Earth. Our ancestors hunted for big game together and took care of each other's sick. They passed down their knowledge in books and stories accessible to strangers. Thanks to our superior capacity for social cooperation, our kind came to dominate the planet's resources and habitats. It wasn't because we're more aggressive or clever than other species. Think of the complex technologies that our cooperative behaviors have made possible. The laptop on which I write these words today is the result of countless people, across bloodlines and generations and continents, contributing their expertise to the mass production and distribution of a powerful tool to millions of others around the world. Even the most aggressive, clever individual working in isolation couldn't have done it.

Cooperative strategies are especially important to defend ourselves against new pathogens because they don't necessarily require high-tech interventions or sophisticated understanding of the pathogen itself to be effective. Even societies with the most rudimentary sense of how pathogens spread can implement effective containment strategies by capitalizing on their ability to work cooperatively. The Acholi people of Uganda are one of the few ethnic groups in Africa whose traditional beliefs about infectious diseases have been studied by medical anthropologists. Many believe that diseases spread through sorcery and spirits. Their traditional responses to epidemics nevertheless limit pathogens' spread: at the first sign of contagion, they work together to isolate the sick, mark their homes with long poles of elephant grass, warn outsiders not to enter affected villages, and refrain from a number of potentially disease-transmitting behaviors, including socializing, sexual intercourse, eating certain foods, and traditional burial practices.[1]

Larger, more formally organized societies can implement even more efficient containment strategies based on cooperative behaviors, like quarantines and those made possible by rapid long-distance communications. They're well poised to do so. Many of the institutions of modern society, after all, are designed to enhance our natural capacities for cooperation, by punishing noncooperators and encouraging the rest of us to pursue even relatively mundane collective actions, like paying taxes and getting flu shots.

And so, when pandemics unfold, it's not just because peculiarly aggressive pathogens have exploited passively oblivious victims or because we've inadvertently provided them with ample transmission opportunities. It's also because our deeply rooted, highly nuanced capacity for cooperative action failed.

In a general sense, this happens when sufficient numbers of individuals choose to pursue their own private interests rather than public ones. For obvious reasons, there are various economic and biological theories that attempt to quantify the conditions under which such choices are made. A simple way of looking at it is by considering the costs and benefits to the individual. The costs of cooperating include the loss of opportunities to pursue private interests (among other things) while the benefits include the increased likelihood of reciprocal behaviors from others and freedom from their opprobrium (among other things). So long as the costs don't exceed the benefits, people will choose to cooperate. Take paying taxes. The cost to me is that I can't use my tax money to buy, say, a new couch. But the benefit is that the government will fund my public library and will not send the IRS out to get me. And so I pay my taxes.[2]

But if the costs of cooperation outweighed the benefits, perhaps I wouldn't. This is what happened on a citywide scale in nineteenth-century New York and is happening on a global scale in many countries today. The factors that allowed cholera to take hold of newly industrializing cities—lack of faith in political governance and the rapid growth of the industrial economy—simultaneously conspired to make selfishness pay off. A bounty of new riches and power tempted private interests, but since the regulatory infrastructure required to restrain their excesses had yet to be built, they faced few penalties when their pursuit of that bounty undermined public health. As the power and influence of private interests eclipsed that of public entities, the strategies that could have obstructed cholera crumbled.

The most blatant example of this in nineteenth-century New York concerns the hijacking of the city's drinking-water supply. It's true, as mentioned earlier, that the island of Manhattan had scarce freshwater supplies, due to the polluted Collect Pond and the brackish Hudson and East Rivers. But the city had another option besides drinking the contaminated groundwater under its tenements and privies: tapping the Bronx River, a freshwater river that flowed from what is today Westchester County twenty-four miles south to empty into the East River.

A physician named Dr. Joseph Browne and the engineer William Weston had proposed that the city build a public waterworks to provide clean, uncontaminated Bronx River water to New Yorkers in 1797. Such a system was affordable: Browne and Weston estimated it would cost the city $200,000, which could be paid for with a new tax. It was technically feasible. Several prominent cities in the industrial world were in the process of building elaborate clean-water distribution systems, such as Philadelphia, where steam engines lifted river water high into reservoirs and then piped it to residents. It would have assured that New Yorkers' drinking water was uncontaminated by fecal bacteria. The Bronx River water flowed upstream of the city and its privies, and Browne and Weston planned to filter its waters through a bed of sand and gravel, in a process now known as "slow-sand filtration." That would have removed more than 90 percent of the bacteria and protozoans in the water.[3]

And it would have substantially improved the quality of life in the city in ways that were obvious to contemporary observers. New Yorkers often complained about the city's lack of water to clean the streets and fight fires. They worried about the health impact of the filthy streets, which according to conventional wisdom endangered the public's health by allowing "pestilential diseases" to fester. ("In suggesting the means of removing the causes of pestilential diseases," a group of leading physicians in New York City reported in 1799, "we consider a plentiful supply of fresh water as one of the most powerful.") And they lived in fear of fire. Fire alarms sounded in 1830s New York City at least once a day. Single conflagrations could level whole neighborhoods of wooden structures. One fire in December 1835 demolished every building south of Wall Street and east of Broad Street, including more than five hundred stores. The public waterworks proposed by Browne and Weston would solve both of these problems.[4]

But private interests in pursuit of ideological power and riches scuttled the plan.

The charming, urbane lawyer Aaron Burr—or, as *The Huffington Post* put it in 2011, the "bad boy of America's founding"—was a New York State senator when Weston and Browne came up with their proposal to tap the Bronx River.[5] Burr was no ideologue, but he was politically ambitious, which meant aligning himself in the dominant political battle of the day: between the Federalists, mostly bankers and businessmen who wanted to strengthen federal institutions in the still-young nation, and the Republicans, who represented the small farmers and other skeptics who opposed them. Despite his patrician background, Burr threw his lot in with the Republicans and figured out a way to strengthen their hand: by creating a new bank.[6]

The Federalist Alexander Hamilton had won a state charter for the bank he'd started, the Bank of New York, in 1791. According to the Republicans, Hamilton's bank discriminated against them. (That may have been true. "Purely political discrimination against applicants for bank services," writes the Hamilton biographer Ron Chernow, "was entirely in keeping with the spirit of the times and the vagueness of the line demarcating business from politics.") A new bank that catered to Republicans could serve as a political counterweight to Hamilton's bank. The trouble was that creating such a bank required obtaining a charter from the state, and chartered companies had to prove that they provided benefits to the public. That would be easier to prove if the new charter was for a private water company–cum–bank rather than just a bank.

But before Burr could obtain a charter for a private water company–bank, he'd have to derail Weston and Browne's proposal for a public waterworks. He tied up the state funds the plan might have tapped. And he told his fellow state legislators that the public waterworks would cost $1 million to build, not the $200,000 that Browne and Weston proposed.[7]

Thus hobbled, Browne and Weston's feasible and lifesaving plan to tap the Bronx River failed to get a charter from the state. As soon as their proposal died, Burr and his allies in the New York City Council applied for and won a charter to launch a private water company and bank: the Manhattan Company.[8] The charter allowed the company to raise $2 million from private investors, ten times more than what Weston and Browne had initially estimated would be required to tap the Bronx and what other cities, such as Baltimore, had raised for their urban waterworks projects. And it allowed the Manhattan Company to use whatever funds it raised that weren't required for the waterworks on any other company business, such as a bank.[9]

Almost immediately upon being organized, the company started downgrading its plans for the waterworks. It decided against using cutting-edge steam engines. It would use horses to pull the pumps instead.[10] It decided against building a one-million-gallon reservoir in the city. It would build a small reservoir that would hold a tiny fraction—just over 0.001 percent—of that volume.[11] It decided against using iron pipes. It would use wooden pipes.[12]

Worse: even though the bank's charter granted exclusive rights to the uncontaminated waters of the Bronx River, and it had more than sufficient financing to pipe that water into the city, the company decided to tap a cheaper, easier source: the filthy, excreta-filled Collect Pond. It did this despite acknowledging, privately, that the Collect's waters were "disgusting" and unfit for human consumption. "The poor

Bronx will be neglected perhaps forever," sniffed a company official in a letter to a relative.[13]

The company's plans outraged New Yorkers. A newspaper correspondent wrote that by distributing the Collect's "fetid" waters, the Manhattan Company would have the blood of thousands on its hands.[14] "It is abominable indeed for the city to be thus trifled with and abused by the company," another resident wrote to a local newspaper.[15] The merchant Nicholas Low called the company "a greater Pestilence than the Yellow fever."[16]

These complaints could have been grounds for revoking the Manhattan Company's charter. Other states routinely revoked the charters of companies that failed to uphold the public good. Ohio, Pennsylvania, and Mississippi had revoked the charters of banks. New York and Massachusetts had revoked the charters of turnpike corporations for not maintaining the roads. But the Manhattan Company's expansive charter made it untouchable. It had been granted its rights and auxiliary powers in perpetuity.[17]

Over the following years, the Manhattan Company spent a mere $172,261 on waterworks for New York City.[18] It devoted the rest of its substantial finances to the bank it opened at 40 Wall Street in 1799. And, indeed, the bank did serve the interests of the Republican elites who'd founded it. DeWitt Clinton, the city's mayor, who also served as the Manhattan Company's director, received a loan of nearly $9,000 (over $150,000 in today's dollars). Burr received $120,000 in loans—nearly as much as the Manhattan Company had spent on its waterworks.[19]

Burr profited politically as well. With his anti-Federalist credentials burnished, in 1801 he ascended to the post of vice president to Republican president Thomas Jefferson.[20] As even his rival Alexander Hamilton had to admit, the Manhattan Company was "a very convenient instrument of profit and influence," albeit "a perfect monster in its principles."[21] (Hamilton's incisive commentary offended Burr, who challenged Hamilton to a duel. On the morning of July 11, 1804, Burr shot and killed Hamilton under the cliffs of the Palisades.)[22]

The Manhattan Company distributed contaminated groundwater to New Yorkers for fifty years, through both the 1832 and 1849 cholera epidemics. The company finally ended its charade of calling itself a water company at the end of the nineteenth century, its checkered past as a waterworks hidden save for its corporate emblem of Oceanus, the Greek god of water, which it retained until the 1950s. Today the company that poisoned New York City with cholera is known as JPMorgan Chase, the largest bank in the United States and the second largest in the world.[23]

The case of the Manhattan Company might have served as a precautionary tale for the other water-deprived cities that cholera stalked. Yet in fact, it was just the opposite: between 1795 and 1800, eighteen private water companies sprang up in Massachusetts. Between 1799 and 1820, twenty-five private water companies opened for business in New York State.[24] In London, five private companies were formed to sell water to urbanites between 1805 and 1811.[25] In nearly every case, the investment required to supply clean water to growing towns and cities overwhelmed private companies' ability to draw profits. Either they ran out of financing or they downsized, readjusting to less ambitious plans that were more profitable—those that distributed easier-to-reach but more likely contaminated waters. Profits were "rarely great enough to induce the directors to build systems adequate to provide all needs," writes the water historian Nelson Manfred Blake.[26] Like the Manhattan Company's rotten system, rather than prevent the spread of cholera, these badly designed, poorly maintained systems helped to more efficiently distribute it.

⊙

Even without the benefit of clean drinking water, collective actions could have averted New York City's cholera epidemics by preventing infected victims from introducing the vibrio into the city in the first place. Political leaders corrupted by ideological and commercial concerns throttled this containment measure, too.

They could have implemented quarantine. The first one had been enacted by Venice in 1374, when the city's gates and ports were shut for forty days to keep out bubonic plague (thus deriving the method's name, from *quarante giorni* or "forty days" in Italian).[27] That was a pretty good containment measure for a pathogen like bubonic plague, which manifests itself in visible pathology in less than forty days. After being held in quarantine for that long, people, ships, and their goods were, as one historian put it, "medically harmless."[28]

By the end of the seventeenth century, all of the major Mediterranean ports in Western Europe had built heavily policed fortresses, called laz-arettos, to hold ships, passengers, and goods in quarantine. To imple-ment similar measures on land, lines of soldiers were positioned in what the French called cordons sanitaire, or sanitary lines. One of the largest— an army of soldiers in a twenty-mile-wide formation that stretched for twelve hundred miles across the Balkans, with orders to shoot on sight any passersby who failed to submit to quarantine—held the plague out of Turkey in the eighteenth century.

Some historians credit the use of quarantines and cordons sanitaire with Europe's ultimate mastery over the plague, which had disappeared by 1850.[29] The historian Pierre Chaunu called it "one of the greatest victories of baroque Europe."[30] The quarantining of ships may have been responsible for the demise of yellow fever epidemics in New York in the first half of the nineteenth century, too, according to the historian John Duffy.[31]

But as international commerce grew over the course of the nine-teenth century, quarantines and cordons sanitaire came to be seen as unreasonably disruptive to trade. Social reformers and free-trade advo-cates sought more openness across borders, not less. Quarantines were "a most unwarrantable tyranny over the merchant," a prominent New York City newspaper railed in 1798.[32] The business losses incurred under quarantine were "calamitous," according to the physician Daniel Drake.[33] "Quarantine is useless," added the British physician Henry Gaulter in 1833, "and the injury it inflicts on the commercial relations and maritime intercourse of the country is an absolute and uncompen-sated evil."[34]

The very idea that infectious diseases spread from person to person—thus making them vulnerable to isolationist measures like quarantines—came to be seen as old-fashioned as well, a "scientific su-perstition," as the nineteenth-century French physician Jean-Baptiste Bouillaud put it, "with which it is to be hoped that we shall soon be finished."[35] Charles Maclean, in 1824, titled his diatribe against quaran-tines, or rather "engines of despotism," *Evils of Quarantine Laws, and Non-Existence of Pestilential Contagion*.[36]

Nineteenth-century medical elites believed that diseases, rather than being contagious, were the result of environmental phenomena, like stinky miasmas and clouds of gas. As New York's resident physician

James R. Manley summed it up in 1832, cholera was "an atmosphere disease . . . carried on the wings of the wind."[37] What would be the point of holding up shipping traffic and the free movement of people if that was the case?[38]

Maintaining this belief required some mental acrobatics, for the reality of contagion was clear enough to see. In rural areas, where people lived far apart and were less likely to pollute each other's drinking waters, diseases like cholera clearly spread sequentially from one sick person to another, moving methodically from household to household, just the same as the quarantinable diseases of old such as plague and smallpox. But most medical elites lived in cities. They tended to discount the experiences of rural peoples, and in any case epidemics looked different in the cities. There, pathogens like cholera spread by both social contact and contaminated waters imbibed by many people at once. The outbreaks unfolded dramatically and simultaneously, just as if everyone had been enveloped by some inescapable cloud of disease, or had been struck by mass poisoning. And if some people fell ill while others didn't, physicians said this was because of their depraved moral condition: drunks, prostitutes, and other disreputable types were more vulnerable than respectable citizens, they explained. (Here, too, inconvenient evidence to the contrary was easy to dismiss. When people in Montreal wrote to newspapers that cholera was attacking the "respectable," disbelieving editors refused to print their letters. Or if a "respectable" person did die of cholera, medical experts argued that he or she must have had some secret vice.)[39]

With medicine and commerce thus ideologically opposed to quarantines, New York City's resolve to effectively enact them had progressively weakened in the years before cholera's arrival. In 1811, the city council and the state legislature relinquished their authority over enforcing quarantines to local health officers at ports. In 1825, they exempted any vessels arriving from Canton or Calcutta from quarantine restrictions. (Just why these two cities were exempted is unclear.)[40] With the implementation of quarantines left to local health officers, enforcement was piecemeal at best. Steerage passengers on incoming vessels might be inspected, but first-class passengers would be allowed through, healthy or not. Motivated individuals could circumvent quarantine by bribing health officers, escaping poorly guarded quarantine centers, or simply fraudulently declaring themselves to be healthy. Ships easily flouted quarantine restrictions. If, say, health officers at ports such as New York City's required that vessels had to spend two to four days in quarantine before docking, captains would simply steer their ships to nearby ports in New Jersey or Throgs Neck where such restrictions didn't hold.[41]

Nonetheless, New York came close to enacting a quarantine against cholera. The governor of New York State had observed cholera's progress across the Atlantic and into Canada during the spring of 1832. Concerned, he dispatched a physician, Dr. Lewis Beck, to conduct a statewide reconnaissance to determine whether the disease posed a risk to the city.

Beck conducted a detailed study, discovering that cholera cases had started breaking out along the path of the Erie Canal and were indeed headed south to New York City, as a modern visualization of his data shows. By contemporary standards, quarantine would have been the appropriate recommendation to the governor. The pattern of cholera cases did seem to "favor the idea that cholera is contagious," Beck admitted.[42]

But that was just an illusion, he went on. In fact, only the immigrants, the poor, and the drunks sickened, and only in the "filthy part of the village." The respectable people who sickened, he explained, had been infected by "eating of a large quantity of green peas" or partaking "immoderately of cucumbers and other vegetables."[43] New York City had nothing to fear and quarantines were unnecessary. "It appears to be abundantly settled," Beck reported, "that cholera cannot be kept out of a country by quarantine laws."[44]

And so cholera traveled unhindered down the waterways toward New York City. Local people girded themselves by doing things like avoiding green and unripe fruit and adopting the upstanding middle-class mores recommended by their physicians: moderation in labor, food, and sex.[45] People in canal towns strung up large pieces of meat on poles to soak up the cholera vapor. Others burned barrels of tar, hoping to rid the air of cholera.[46]

⊙

The third containment measure that New York City failed to implement continues to falter today: prompt public alerts regarding the arrival and the spread of the disease.

For fear of disrupting trade, the mayor of New York and the city's board of health refused to notify the public about the contagion in their midst. So did every other town and city that was stricken by cholera, which issued vague reports of "sudden deaths" of "unknown disease" rather than admit that cholera had erupted in their community. (Unaffected neighboring towns were less reluctant to call the disease by its name, which is how news about cholera's spread got out.)[47]

Prominent New York City physicians, deluged with patients sick with cholera in the summer of 1832, begged the mayor to issue a public alert. Both he and the board of health denied cholera had emerged at all.[48] Outraged and alarmed at the "tardy and obstinate" behavior of city officials, a group of leading physicians published a blistering public bulletin condemning the municipality (which they called the "Corporation") for valuing "dollars and cents above the lives of the community."

> This no doubt has led them so pertinaciously to deny the existence of Cholera in this city—even after the fact was established by the united testimony of the whole Profession . . . We appeal to the good sense of our Citizens whether there can be any apology for the criminal neglect of the Corporation, in relation to the distresses of the thousands who are now crying out to us for aid . . . It is time that you should be removed from the office, which, instead of dignifying, you only disgrace.[49]

City officials may have destroyed evidence of the arrival of cholera-infected ships in the weeks before the outbreak, too. Following up on claims made by the port physician that the city had secretly quarantined passengers from a cholera-infected ship, investigators found that otherwise intact quarantine-hospital records for the months in question—April, May, and June 1832—had disappeared.[50]

◉

To be fair, the choices that nineteenth-century leaders had to make about whether or not to implement disease control strategies were not between two equally compelling options. The choices were between predictable costs and unpredictable benefits. They knew that quarantines and alerting the public about cholera would disrupt private interests, but they couldn't be sure that either strategy would actually protect the public. It's not surprising, then, that they opted for near-certain private benefits rather than mostly uncertain public ones. Plus they were under no obligation to do otherwise.

By the twentieth century, that had changed. Starting in 1851, about a dozen European nations in conjunction with Russia launched a series of international meetings to hammer out an agreement to alert one another about the presence of infectious diseases within their borders. After five decades of acrimonious debate, by 1903 they'd agreed to report cases of cholera and plague to each other, to implement maritime quarantines for cholera, and to allow other nations to inspect ships from cholera-infected ports, as part of an International Sanitary Convention.

But powerful private interests continued to sabotage their efforts, despite the international agreement. One of the most daring and well-coordinated international conspiracies to conceal an infectious disease outbreak occurred just a handful of years after the convention was signed.

☉

Cholera broke out in Naples, Italy, in 1911. It was the eve of a nation-wide celebration of the country's fiftieth anniversary, which was expected to draw millions of tourists. The Italian prime minister, more interested in protecting commerce and prestige than the health of his people, made his intention to flout the International Sanitary Convention clear in a telegram sent to his public-health authorities: "The aim is to obtain and maintain the greatest possible secrecy" about the unfolding cholera epidemic in the country, he instructed. "The Government will be merciless with all those who are slow or negligent."

Italian authorities paid newspapers and reporters secret monthly retainers of 50–150 lire a month to avoid mention of the dreaded "c" word; they intercepted and censored telegrams that contained the word "cholera"; they tapped the phones of those who might leak the news, threatening them with imprisonment. They conducted nighttime raids on medical societies, confiscating cholera-education materials. And while they continued to maintain records on each case that occurred, they stamped case reports with the word "secret," in bold type, and the reminder: "N.B. No official bulletin was published." Cholera victims were transported to hospitals in the dead of night, while local papers blared, "There is no cholera, and never was!"

   U.S. officials signed on to the cover-up as well. "No unnecessary pub-
licity will be given as to the existence of cholera in Italy," the secretary
of state reassured skittish Italian authorities by telegram. So long as the
Italians promised to take care of the messy cholera discreetly, the United
States would ignore the strictures of the International Sanitary Convention. "Bills of health, after completion, shall be given to the Master of the
vessel, sealed," the secretary of state affirmed, "and knowledge of contents
known only to the consul and the medical officer, not even the Master
is to know the contents." While the surgeon general didn't bother warning the public about the risks of travel to cholera-stricken Italy, in private
correspondence he advised his personal acquaintances to cancel their
plans to visit Italy that summer. The French government also agreed to
the terms of Italy's conspiracy.[51]

   The historian Frank Snowden estimates that the secret cholera epidemic in Italy killed up to eighteen thousand between 1910 and 1912,
and spread into both France and Spain. While the details of Italy's secret
cholera epidemic did not make an appearance in the historical literature
until decades later, when it was exposed in detail by Snowden, close
readers of German novels could have figured it out at the time. The
German novelist Thomas Mann and his wife had visited Italy during
the secret cholera epidemic. In 1912, Mann published his novella *Death
in Venice*, in which a German writer visits Venice, finding the city under some kind of "nameless horror." The writer ultimately perishes after
consuming overripe strawberries, commonly believed by Mann's contemporaries to be a major risk factor for cholera infection.

   Italy's cover-up was among the most daring, but it was hardly the
last. Political leaders continue to prioritize commerce and national
reputation over public health. In 2002, Chinese authorities treated the
emergence of SARS as an official state secret. A spokesman for the
Guangdong health department said that information about the brewing
epidemic would be dispensed solely by "the party propaganda unit" and
that any physician or journalist who reported on the disease risked persecution, the critic Mike Davis reported. Aside from a few newspaper
reports from the city of Foshan—which mentioned only a spate of unex-
plained respiratory deaths—the international community and public-
health authorities around the world had no idea about the outbreak.[52]

   It was only months later, when a local resident happened to mention what was happening in Guangzhou in a message to an online

acquaintance, that the international community learned of the new pathogen's emergence. The recipient of the message forwarded it to a retired navy captain named Dr. Stephen Cunnion, who posted the following query to an infectious-disease reporting system run by an international medical society, the Program for Monitoring Emerging Diseases (Pro-MED), on February 10, 2003. "This morning I received this email," he began, "and then searched your archives and found nothing that pertained to it. Does anyone know anything about this problem? 'Have you heard of an epidemic in Guangzhou? An acquaintance of mine from a teacher's chat room lives there and reports that hospitals are closed and people are dying.'"[53]

After news of the outbreak reached the Beijing office of the World Health Organization, Chinese officials continued to stonewall. They admitted to only a few deaths from "atypical pneumonia." They blocked, at least initially, investigative teams from the WHO from inspecting the military hospitals where SARS patients were being treated. It was only after a sufficiently alarmed WHO advised visitors to stay away from Hong Kong and Guangdong that the Chinese health minister publicly acknowledged the presence of the new killer virus. Even then, the minister insisted that the pathogen had been contained and that south China was safe, neither of which turned out to be true.[54]

The government of Cuba similarly suppressed news about a 2012 outbreak of cholera. According to *The Miami Herald*, Cuban authorities told local doctors to list cholera deaths as being caused by "acute respiratory insufficiency." "We have been forbidden from using the word cholera," a local man told the newspaper, adding that people had already been arrested and detained for transgressions. While news of cholera's continuing spread leaked out of the country, the government announced that the cholera outbreak had been contained. In December 2012, they detained a journalist who'd reported on the cholera outbreak. (The doctor who had publicly reported on the country's 2000 dengue fever outbreak had been imprisoned for over a year.)[55] When you report cholera, "you get into trouble," as a government official in Dar es Salaam, Tanzania, told the journalist Rose George.[56]

The Saudi Arabian government attempted to silence the virologist who discovered a novel coronavirus that had first popped up in a patient in a hospital in Jeddah in the fall of 2012. Cognizant of the SARS-like

threat that the new virus posed, a virologist at the hospital, Dr. Ali Mo-
hamed Zaki, posted his findings on Pro-MED, alerting its sixty thousand worldwide subscribers. By all accounts, Zaki's timely warning averted
what could have become a global outbreak. The coronavirus was rapidly sequenced, diagnostic tests devised, and public-health authorities
around the globe discovered over a hundred more victims of what came
to be called Middle East respiratory syndrome (MERS). According to
Zaki, the Saudi Arabian ministry of health was not pleased. "They were
very aggressive with me," he says. "They sent a team to investigate me . . .
And now they force the hospital administration to force me to resign."
The man who may have prevented a pandemic lost his job and had to
relocate to Egypt.[57]

It's not only governments with reputations for repression that have
squashed information about new pathogens. The democratically elected
government of India attempted to suppress word of NDM-1, too. The
first reports on NDM-1 and its spread through India's medical tourism
industry appeared in the international medical literature in August
2010, in a paper in *The Lancet* coauthored by British and Indian scientists. Immediately upon its release, Indian medical tourism advocates
started denying the public-health significance of NDM-1. "Such super-
bugs are everywhere," pooh-poohed Dr. Vishwa Katoch, the Indian government's secretary of health research. "India has a problem like other
countries." Research on NDM-1 and the naming of the plasmid after
the city of New Delhi where it was first isolated was "a conspiracy to
hurt Indian medical tourism," opined *The Indian Express*. The conclusions of the research on NDM-1—that the business of medical tourism
might need to be restrained—were "unfair and scary," added *The Hindu*
newspaper.[58]

Indian government authorities cracked down on the Indian scientists who had been involved in the NDM-1 study, intimating in letters
and private meetings that their research on the new pathogen broke the
law. "Permission is required to be obtained by the relevant authority," a
letter to the researchers from the health ministry stated. "You are hereby
asked to explain the details about the study conducted." University of
Cardiff's Timothy Walsh, who spearheaded the research, was accused
of being a spy and was deluged with hate mail, he says. According to the
Indian government, Walsh says, "I'm the devil incarnate and eat babies

for breakfast." With international collaboration on NDM-1 sharply curtailed by the Indian government's interference, Walsh was forced to enlist journalists to acquire samples in India to continue his research on the plasmid.[59]

⊙

The nineteenth century is infamous for the rise of the unethical capitalists derided as "robber barons," but it's post-twentieth-century globalization that has concentrated unprecedented power in the hands of private interests. Of the one hundred largest economies in the world, only forty-nine are countries; fifty-one are private corporations.[60] By 2016, the richest 1 percent of people in the world will control more than half of the planet's total wealth.[61]

The influence of these private interests dwarfs that of the public institutions that might seek to regulate them. And so when private interests run contrary to those of public health, it's often public health that gets the shaft. A good example of this is in the area of antibiotic consumption.

The fact that wanton consumption of antibiotics—using either more or less than the precise amount required to tame an infection—leads to the development of antibiotic-resistant pathogens has been long known. It was first outlined by Alexander Fleming, the scientist who discovered penicillin. "I would like to sound one note of warning," he said, in his speech accepting the Nobel Prize in Physiology or Medicine in 1945. "It is not difficult to make microbes resistant to penicillin in the laboratory by exposing them to concentrations not sufficient to kill them, and the same thing has occasionally happened in the body. The time may come," he went on, presciently,

> when penicillin can be bought by anyone in the shops. Then there is the danger that the ignorant man may easily underdose himself and by exposing his microbes to non-lethal quantities of the drug make them resistant. Here is a hypothetical illustration. Mr. X has a sore throat. He buys some penicillin and gives himself, not enough to kill the streptococci but enough to educate them to resist penicillin. He then infects his wife. Mrs. X gets pneumonia and is treated with penicillin. As the streptococci are now resistant to penicillin the treatment fails. Mrs. X

dies. Who is primarily responsible for Mrs. X's death? Why Mr. X whose negligent use of penicillin changed the nature of the microbe.[62]

While Fleming warned about the perils of underuse of antibiotics, the same risks apply with overuse. But while judicious use of antibiotics served the needs of public health, rampant consumption served those of private interests. In many countries, hospital physicians found it convenient to dose whole wards with antibiotics, indiscriminately. Patients found comfort in consuming antibiotics for colds and flus and other viral infections, for which they are useless. Farmers profited by giving antibiotics to their livestock, which for reasons that are still unclear made them grow faster and helped them thrive in factory farms. (Their provision of low-dose antibiotics to their livestock for "growth promotion" accounts for 80 percent of all antibiotic consumption in the United States.) Cosmetic companies enlarged their markets by packaging antibiotics in their soaps and hand lotions.[63] By 2009, the people and animals of the United States were consuming upward of 35 million pounds of antibiotics annually.[64] "Fleming's warning," one microbiologist writes, "has fallen on ears deafened by the sound of falling money."[65]

In countries like India, where there are fewer restrictions on antibiotic consumption, overuse is rampant. Even the most high-end antibiotics are available without a prescription. The poor, who can't afford full courses of the drugs, pop one or two tablets at a time, modern-day Mr. X's. While hundreds of thousands of Indians die every year for lack of access to the right antibiotics at the right time, antibiotics are routinely used by others for nonbacterial conditions, such as colds and diarrhea. Studies suggest that up to 80 percent of patients with respiratory infections and diarrhea in India—conditions unlikely to be alleviated by the use of antibiotics—are given antibiotics. Precise diagnostics that would rule out this risky and ineffective use are expensive and hard to come by. Plus, the pharmacists who fill scrips for antibiotics make a good living on it, as do the companies that sell the drugs.[66]

Antibiotics, had they been well stewarded, experts say, could have effectively treated infections for hundreds of years. Instead, one by one our bacterial pathogens have figured out how to rout the onslaught of antibiotics to which they've been indiscriminately subjected. We now face what some experts call an era of "untreatable infections." Already, a "growing minority" of infections have become "technically untreatable," as David Livermore of the U.K.'s national Antibiotic Resistance Monitoring lab wrote in 2009.[67]

Controlling antibiotic consumption would almost certainly solve the problem. Places that use antibiotics sparingly whether due to the difficulty of accessing them, like Gambia, or due to more conscious restraints, like Scandinavia, experience low rates of drug-resistant microbes. MRSA is rare in Finland, Norway, and Denmark, as well as the Netherlands, even in hospitals. Since 1998, new patients admitted to Dutch hospitals have been swabbed and tested for MRSA, and if they're found positive, they're treated with antibiotics and kept in isolation until they've demonstrably shaken the bug. By 2000, only 1 percent of staph strains in Dutch hospitals showed resistance to methicillin and its cousin compounds. In Denmark, national guidelines restricted antibiotic prescriptions, and MRSA rates dropped from 18 percent of all staph in the late 1960s to just 1 percent within ten years.[68]

But even as the toll of drug-resistant bacteria rises, vested interests are reluctant to admit there's a problem, and weak public institutions even more reluctant to challenge them. Attempts to slow the consumption of antibiotics in the United States—and threaten the financial interests of the livestock and the drug industries as well as doctors and hospitals—have floundered again and again.

In 1977, the FDA proposed removing the antibiotics penicillin and tetracycline from use in livestock for growth promotion, but Congress blocked the move. Then in 2002, the FDA announced that it would regulate the use of antibiotics in livestock only if the practice could be proved to cause high levels of drug-resistant infections in people. Even experts who believe it does admit that the connection is nearly impossible to conclusively prove. Finally, in response to a lawsuit filed by a coalition of NGOs, in 2012 a federal court ordered the FDA to regulate the practice anyway.[69] In December 2013, the FDA issued a set of voluntary guidelines on antibiotic use in livestock, but it was so full of loopholes that one activist fighting for stricter controls called it "an early holiday gift to industry."[70]

The government has been similarly recalcitrant to rein in antibiotic use in hospitals and doctors' offices. In 2006, after a fractious ten-year-long effort, the CDC issued voluntary guidelines on how to prevent drug-resistant bacteria spread in hospitals. The guidelines were so jumbled, the Government Accountability Office reported, that they "hindered efforts" to put them into useful action, as the journalist Maryn McKenna recounts in her history of MRSA.[71]

Finally, in September 2014, the White House issued a series of guidelines on the issue. Whether political leaders had finally challenged commercial interests still remained unclear. The guidelines fell roughly into two categories: those that would restrict the use of antibiotics—and directly conflict with the interests of drug companies, farmers, and hospitals—and those that would foster the development of new antibiotics and diagnostic tests to take the place of the old. Suggestively, the former were delayed, while the latter fast-tracked. The implementation of the guidelines that would restrict consumption was put off until 2020, pending the actions of a new advisory council and task force. But the government immediately announced plans to provide a windfall to the pharmaceutical industry, with a $20 million prize for the development of a rapid diagnostic test to identify highly antibiotic-resistant bacteria.[72]

The burden of drug-resistant pathogens extends beyond the people who will die from infections for which no effective treatment exists. A much larger group of people will suffer infections for which only a select few antibiotics will work. They will show up at hospitals and clinics with what seem to be routine infections and be erroneously treated with the wrong antibiotics. Studies suggest that anywhere from 30 to 100 percent of patients with MRSA are initially treated with ineffective antibiotics.[73] Delays in effective treatment allow the pathogen to progress until it is too late. A simple urinary tract infection, for example, becomes a much more serious kidney infection. A kidney infection becomes a life-threatening bloodstream infection.[74]

And then there are those of us like me and my son. It used to be that staph infections didn't really affect otherwise healthy people like us. It was a problem for people weakened by hospitalization, or acute rehab units, or long-term care facilities. But then in 1999, *Staphylococcus aureus*, in response to an onslaught of antibiotics, spawned a drug-resistant form, picked up the ability to secrete a toxin, and escaped from the American hospitals in which it first emerged. By 2001, 8 percent of the U.S. general population was colonized by MRSA bacteria, mostly inside their noses.[75] Had the surveyors sampled more obscure bodily locations, that number might have been even larger. Two years later, 17.2 percent were colonized. It's not just the skin and soft-tissue infections that MRSA most often causes in healthy people. If, via a wound, say, or a dental procedure, or a botched lancing of a boil, MRSA penetrates deeper into the body, the consequences are dire. Tissue-destroying infections of the lung (necrotizing pneumonia) and flesh-eating disease (necrotizing fasciitis) are just two of the unpleasant—and often deadly—possibilities. By 2005, MRSA had caused more than 1.3 million infections in the United States, causing what experts have called a public-health crisis in the nation's emergency rooms and doctors' offices.[76]

For now, the MRSA strain most commonly picked up by people outside of hospitals, the USA300 strain, while impervious to penicillin and other similar "beta-lactam" antibiotics, is still susceptible to non-beta-lactam antibiotics. That may not help much if you have necrotizing pneumonia—38 percent of patients die within forty-eight hours of being admitted to the hospital—but it is still something.[77] But not for long, perhaps. Staph strains that can resist non-beta-lactam drugs as well have already been spotted.[78]

There are few new drugs on the horizon. Since antibiotics are not used for very long, there's little market incentive for drug companies to develop new ones. The market value of a brand-new antibiotic is just $50 million, a paltry sum for a drug company considering the research and development costs incurred to create such drugs. As a result, between 1998 and 2008, the FDA approved just thirteen new antibiotics, only three of which boasted new mechanisms of action.[79] In 2009, according to the Infectious Diseases Society of America, only sixteen of the hundreds of new drugs in development were antibiotics. None targeted the most resistant and least treatable gram-negative bacteria, like those endowed with NDM-1.[80]

⊙

The U.S. government is not alone in falling prey to the rising power of private interests and allowing pathogens to spread as a result. It's happened to our premier international agency, the World Health Organization, as well.

The agency was created by the United Nations in 1948 to coordinate campaigns to protect global public health, using dues collected from the UN's member nations. But over the course of the 1980s and early 1990s, the major donor nations, skeptical of the UN system, started to slowly starve it of public financing. (They introduced a policy of zero real growth to the UN budget in 1980, and of zero nominal growth in 1993.)[81] To make up the budgetary shortfall, the WHO started to turn to private finance, collecting so-called voluntary contributions from private philanthropies, companies, and NGOs, as well as donor countries. In 1970, these voluntary contributions accounted for a quarter of the agency's budget. By 2015, they made up more than three-quarters of the agency's nearly $4 billion budget.

If these voluntary donations simply substituted for missing public funding, they wouldn't make much of a difference in the way the WHO functions. But they don't. Public funding (via annual dues from member nations) doesn't come with any strings attached. The dues are simply assessed and collected, and the WHO is in charge of deciding how to spend the money. That's not true of voluntary contributions. By making a voluntary contribution, individual donors buy control at the WHO. They can bypass the WHO's priorities and allot the money for whatever specific purpose they like.[82]

Thus the WHO's activities, the agency's director-general Margaret Chan admitted in an interview with *The New York Times*, are no longer driven by global health priorities but rather by donor interests.[83] And those interests have introduced a pronounced distortion into the WHO's activities. While the agency's regular budget is allocated to different health campaigns in proportion to their global health burden, according to an analysis of the agency's 2004–2005 budget, 91 percent of the WHO's voluntary contributions were earmarked for diseases that account for just 8 percent of global mortality.[84]

Many of the WHO's deliberations are conducted behind closed doors, so the full extent of the influence of private donors is unclear. But their conflicts of interest are plain enough. For example, insecticide manufacturers help the WHO set malaria policy, even though their market for antimalarial insecticides would vanish if malaria actually receded. Drug companies help the WHO determine access-to-medicine policies, despite the fact that they stand to lose billions from the cheaper generic drugs that would improve patients' ability to get the treatments they need. Processed food and drinks companies help the agency craft new initiatives on obesity and noncommunicable diseases, even though their financial health depends on selling products that are known to contribute to these very problems.[85]

As the integrity of the WHO has been degraded by private interests, so has its ability to effectively lead global responses to public-health challenges. During the Ebola epidemic in West Africa in 2014, the weakened agency was unable to muster a prompt response. It turns out one reason why is that the agency had been forced to compromise on the integrity of the officials they'd hired. Rather than being appointed for their commitment to global health, they'd been appointed for political reasons. When the affected countries wanted to downplay the epidemic so as not to upset their mining companies and other investors, the WHO's politically appointed local officials went along with it. As an internal document leaked to the Associated Press revealed, they refused to acknowledge the epidemic until it was too late to contain. They failed to send reports on Ebola to WHO headquarters. The WHO official in Guinea refused to get visas for Ebola experts to visit the afflicted country. It wasn't exactly a cover-up, but the agency's top polio official, Bruce Aylward, admitted in the fall of 2014 that WHO's actions ended up "compromising" the effort to control the Ebola epidemic rather than aiding it.[86]

As the effectiveness of the WHO's leadership wanes, the influence of private global health outfits grows. Some have started to eclipse public ones like the WHO entirely. Bill Gates, the cofounder of the computer giant Microsoft, used the fortune he accumulated from the global high-tech economy to form the Bill and Melinda Gates Foundation, the world's largest private philanthropy, in 2000. The Gates Foundation soon became the world's third-biggest financier of global health research, outstripped only by the U.S. and U.K. governments, and one of the world's single largest donors to the WHO.[87] Today it's the privately run Gates Foundation that sets the global health agenda, not the WHO. In 2007, the foundation announced that resources should be devoted to the eradication of malaria, contrary to a long-established consensus among scientists in and outside the WHO that controlling the disease was safer and more feasible. Nevertheless, the WHO immediately adopted the Gates plan. When the agency's malaria director, Arata Kochi, dared to publicly question it, he was promptly put on "gardening leave," as one malaria scientist put it, never to be heard from again.[88]

The well-intentioned people at the Gates Foundation have no particular private interest that directly conflicts with their ability to promote global health campaigns in the public interest, or at least none that we know of.[89] But if they did, there'd be no mechanism to hold them accountable for it. Powerful private interests unfettered by public controls, even when they have charitable intentions, are like royalty. We've ceded control to them and now we must simply hope that they are good. Our ability to mount a cooperative defense against the next pandemic depends on it.

⊙

Of course, even if political leaders are corrupt and political institutions are rotten, people can still cooperate with each other. They can take matters into their own hands, launching their own cooperative efforts to contain pathogens. For example, when city leaders failed to alert New Yorkers to the spread of cholera in the nineteenth century, private physicians banded together and issued their own bulletins.

Such actions make sense. And extreme events do tend to bring people closer together. Think of New Yorkers after the September 11 terror attacks or in the wake of recent hurricanes. But that's not what tends to happen when pandemic-causing pathogens strike.

Unlike acts of war or catastrophic storms, pandemic-causing pathogens don't build trust and facilitate cooperative defenses. On the contrary, due to the peculiar psychic experience of new pathogens, they're more likely to breed suspicion and mistrust among us, destroying social bonds as surely as they destroy bodies.

SIX

BLAME

People on the street eye us warily as my guide and I slowly drive the smooth, wide roads of Cité Soleil, a seaside slum along the outskirts of Port-au-Prince. The dusty, flat neighborhood is mostly treeless, the sun beating down on its shacks and crumbling, bullet-ridden buildings. It's around noon on a weekday, but despite high unemployment in the slum, the streets are empty. I'd come to visit Cité Soleil in the summer of 2013 to get a sense of how the people most vulnerable to the cholera epidemic felt about it. But I'm reluctant to approach the few people we see, sitting on overturned buckets in a patch of shade or ambling in the expanse of packed dirt in front of their shacks. They frown at us as we pass, although whether that's because of the sun in their eyes or something else it's hard to say.

The sense of incipient violence grows more palpable as we continue on to the edge of the neighborhood, where the city dump is located. Here, people who make their living scavenging through Port-au-Prince's garbage are walking around, and we can see clumps of people talking in the distance up the road. They seem approachable to me, but before we can reach them, we're stopped by the helmeted guards posted in front of the gates of the dump. We're not allowed to wander around on our own, they tell us sternly, without a uniformed official with us. Without visible proof of government permission, someone may "act crazy," they say. This makes no sense to us, and one of the guards has what looks like an empty ChapStick container shoved up one of his nostrils, which somewhat undermines his authority. We reluctantly climb back into the car anyway. When I snap a few photos before ducking in, he angrily raps on the window. If anyone had seen what I just did, he scolds, they'd throw rocks at me or worse.

We found people to talk to at the garbage-strewn shore. A group of about a dozen young men milled around aimlessly, hoping for work on one of the small wooden boats moored nearby. They surrounded us as soon as we parked the car. The knot of fishermen untangling ropes in a bombed-out cement structure wouldn't talk to us, or allow us to take their photographs, but the young men were open to talking. Still, after a few minutes, they told us we had to leave. Somewhere deep inside the neighborhood a conflict was brewing. As we made our way out of the slum, two shiny white trucks emblazoned with the letters UN pulled over in front of us, emitting a stream of soldiers in full combat gear. They trotted off purposely into the interior, single file, rifles in hand.

The details of this specific mobilization were unclear, but I knew that the ongoing cholera epidemic had inflamed a history of violent clashes between local people and outsiders—specifically, United Nations peacekeeping troops. The troops had first arrived in the country in 2004. In principle, their mission was aimed at maintaining peace and order in Haiti, but most Haitians understood the UN presence as an occupation that took over where U.S. troops, which had been sent to Haiti three times over the last century, left off. (The U.S. ambassador admitted as much in a leaked 2008 cable, recognizing the UN troops as an "indispensable tool in realizing USG [U.S. government] policy interests in Haiti.") Those interests, since the 1990s, were primarily to suppress militants, or what U.S. leaders called "criminal gang members," who supported the liberation theologian and deposed Haitian leader Jean-Bertrand Aristide. Cité Soleil was a stronghold of both militancy and crime.

As a result, there hadn't been much peaceable about the UN's activities in Haiti. Between 2004 and 2006, for instance, UN troops helped the Haitian police and paramilitary forces kill an estimated three thousand people and imprison thousands of Aristide supporters.[1] A Haitian parliamentarian called the UN troops "a fish bone stuck in our throats."[2]

The cholera epidemic had sparked another outburst of violence between locals and UN soldiers. In Saint-Marc, a crowd threw stones at a

local cholera treatment center as UN troops fired upon them. Elsewhere, at a Red Cross clinic, stone-throwing students were confronted
by rifle-wielding soldiers. In Port-au-Prince, mobs tore down tents set up
to treat cholera patients.[3] The situation in Cap Haitien was so dire, with
mobs burning down police stations, that the entire city was put under
lockdown. Schools, shops, and offices closed while aid workers hid in
office buildings, the walls covered in graffiti reading UN=KOLERA.
Cholera had ignited such discord and violence that the UN's humanitarian chief called the disease "a threat to national security."[4]

⊙

Cholera riots date back to the nineteenth century. Across Europe and
the United States, paroxysms of violence fanned out in cholera's wake, a
"pandemic of hate," as the historian Samuel Cohn has described it, that
nipped on the heels of disease like an angry dog.[5]

It doesn't really make sense, on the face of it. It would seem that in
times of social stress—say, the arrival of a deadly contagion—the appropri-
ately healthful response would be to move even closer together, to clasp
hands and stand shoulder to shoulder in the face of the intruder. Instead, epidemics of new disease often set in motion "an inexorable
collapse of morals and manners," as the critic Susan Sontag wrote.[6] They
"spawn sinister connotations," the medical historian Roy Porter added.[7]
And the discord sparked by epidemics isn't generalized or diffuse. As in
Haiti, it's often focused with laser-like intensity on specific groups of
people—scapegoats—who, among all the potentially culpable groups and
social factors, are fingered as being especially responsible for the epidemic.

Calling the troops scapegoats is not to say that there was no basis
for the animosity between UN soldiers and local Haitians, or that UN
soldiers were not complicit in cholera's spread. Indeed, the UN had
hired soldiers from cholera-struck Nepal, whom they could pay a fraction
of what the United States paid its own soldiers, and it was these soldiers
who had introduced cholera into Haiti. But while the troops introduced
the pathogen, they could not logically be held responsible for the way it
ignited across the country. That had to do with larger, more deeply
rooted problems beyond their immediate control, like poverty, the lack of
clean water, and the dislocations caused by the preceding earthquake. Nor
were the troops, at the time when they were attacked, actively contributing

to the epidemic. On the contrary, they and others associated with them were ostensibly trying to help.[8]

A handful of psychological studies offer some clues about the social and political contexts in which scapegoating is most likely to occur. These studies attempt to measure subjects' willingness to blame scapegoats under various experimental conditions. In one study, subjects who were reminded of their powerlessness over a social crisis or the inability of their government to protect them from it expressed a greater desire to punish a scapegoat, compared to subjects who were simply told of the presence of the crisis. Other subjects, who were reminded of their own role in contributing to a crisis, expressed a similar eagerness to punish scapegoats.[9] In another study, subjects who perceived less control over their lives believed that scapegoated groups were more powerful than did subjects who perceived more control over their lives.[10] Just who the scapegoat is makes a difference, too. Groups that seem incompetent, weak, or circumscribed in their social power are less likely to attract blame. It's groups that seem feasibly complicit in social crises (a corporation as opposed to the Amish, in the case of environmental damage, for example), powerful, and yet also mysterious who are the most likely targets, studies have found.[11]

The psychiatrist Neel Burton, who has written about the psychology of scapegoating, sees it as a form of projection. Powerlessness and complicity, he says, are uncomfortable feelings that people naturally seek to expunge or escape, and one way to do that is to project them onto others. When those others are punished, the old feelings of powerlessness and guilt are transformed into feelings of mastery or even "piety and self-righteous indignation."[12]

That may be why epidemics caused by novel pathogens so often lead to violent scapegoating. Because they're poorly understood and specialize in striking societies with weak and corrupt social institutions, such epidemics are especially adept at disrupting people's sense of control over their environment. At the same time, their ravages are not inescapable, either, like the effects of wars or floods. Certain people are struck, while others are not, suggesting some form of complicity, however opaque.

Ancient peoples captured the impulse to scapegoat during social crises in telling rituals. In ancient Greece, during epidemics or other social crises, beggars or criminals called the "pharmakos" were ritually stoned, beaten, and driven out of society. In ancient Syria, female goats designated as the vehicle of evil were decorated with silver and driven out into the wastelands to die alone during royal weddings. The word "scapegoat" itself derives from a ritual described in Leviticus in the Old Testament, in which God commands Aaron to sacrifice two goats for the Day of Atonement. One was to be slaughtered. The other goat, for "Azazel," was to be symbolically laden with all of the transgressions of the Israelites and then sent out into the desert to perish alone. The ritual sacrifice of the Azazel goat, which the King James version of the Bible translates as "scapegoat," dramatized people's desire to expunge the powerlessness and guilt of life in a world of capricious hazards, from famines to epidemics.[13]

☉

Scapegoating is particularly disruptive during epidemics because it often targets the very groups of people most likely to be able to contain epidemics and alleviate their burden.

During the nineteenth century, it was physicians and religious leaders who were often the targets for violence. When cholera hit Europe in 1832, rumors that hospitals were in the business of killing patients to rid society of those deemed "surplus" made the rounds. People stoned and assaulted local physicians, accusing them of killing cholera victims for the express purpose of dissecting their bodies. More than thirty riots erupted in Britain and France between February and November 1832, from stone-throwing episodes that contemporaries called "petty tumults" to melees involving hundreds.[14]

During cholera outbreaks in New York, mobs attacked quarantine centers and cholera hospitals, and blocked health officials from removing cholera-struck corpses from their tenements. (During one confrontation, health officials were forced to lower a coffin to the ground via a window.)[15] In cholera-struck Madrid in 1834, citizens became convinced that the monks and friars—who contentiously supported the king's brother's designs upon the throne—had brought the cholera by poisoning the wells. Angry mobs converged in Madrid's public squares, ransacking religious houses and Jesuit church buildings and murdering fourteen priests. The Franciscans of San Francisco suffered heavily: forty were stabbed, drowned in wells, hanged, or hurled from rooftops. "The bloody scenes did not end until well into the night," the historian William J. Callahan notes.[16]

Immigrants were similarly targeted for violent scapegoating. Like health-care workers and religious leaders, they were considered somehow complicit in outbreaks: the correlation between immigrant neighborhoods and the prevalence of disease was clear enough to see. Of course, the building owners who crowded immigrants into urban tenements and the commercial interests that dominated trade and travel routes contributed as much if not more to the spread of disease, in relatively easy-to-see ways, too, but were spared from the violence. Immigrants, with their mysterious culture and outsider status, suffered the brunt instead.[17]

Cholera's arrival led once welcoming communities to refuse to rent rooms to passing immigrants and travelers. "Distressed strangers were obliged to sleep in the streets, in the fields," and in beds "made principally with bed clothes, and a few boards and sticks," a local newspaper in Lexington, Kentucky, noted in 1832.[18] Residents of the towns lining the Erie Canal refused to let boats enter their waters, or to let anyone on passing boats disembark, even passengers attempting to return home.[19]

The particular immigrant groups blamed for cholera's spread varied over the decades. In the 1830s and 1840s, it was the Irish. "Being exceedingly dirty in their habits, much addicted to intemperance and crowded together in the worst portions of the city," the New York City board of health noted in 1832, the "low" Irish "suffered the most" from cholera. The Irish "brought the cholera this year," Philip Hone complained in his diary, "and they will always bring wretchedness and want."[20] In 1832, fifty-seven Irish immigrants living in an isolated clearing in the woods of Pennsylvania—they had been hired to clear a path for a new rail line between Philadelphia and Pittsburgh—were quarantined and then secretly massacred, their shacks and personal belongings burned to the ground. "All were intemperate, and ALL ARE DEAD!" local papers gleefully reported.[21] Investigators unearthed the workers' smashed and bullet-riddled skulls from a mass grave in 2009.[22]

In the 1850s, the wave of violence that followed cholera crashed upon Muslims, in particular, pilgrims on Hajj. Muslim religious stricture requires that all practitioners perform the Hajj pilgrimage to Arafat, about twelve miles east of the Saudi Arabian city of Mecca, at least once in their lives.[23] As the pace of international trade and shipping picked up, so did the number of Hajjis. In 1831, 112,000 pilgrims participated in Hajj; by 1910, an estimated 300,000 did.[24] Cholera outbreaks followed as well. In one of the worst outbreaks, in 1865, cholera killed fifteen thousand Hajj pilgrims.[25]

Anxieties among Western elites that the Hajj would infect the West with cholera—which they continued to describe as a disease of Asian filth despite its proven affinity for Western cities—grew accordingly. The series of international meetings convened between 1851 and 1938, which resulted in the 1903 International Sanitary Convention and served as a precursor to the World Health Organization, focused specifically on how to selectively confine Meccan filth from contaminating Western society. "Mecca, I hold, is the place of danger for Europe," the British doctor W. J. Simpson put it, "a perpetual menace to the Western world." Indeed, added another influential Brit, "the squalid army of Jagganath with its rags and hair and skin freighted with infection may any year slay thousands of the most talented and beautiful of our age in Vienna, London, or Washington."[26] The trouble with Hajj pilgrims from India, another added, is that they "care little for life or death," but "their carelessness imperils lives far more valuable than their own."[27] The French recommended sealing off the Middle East entirely, by selectively banning Hajj pilgrims from traveling by sea at all, forcing them to journey to Mecca in caravans over the desert.[28]

In the 1890s, cholera-fueled scorn in New York City fell upon Eastern European immigrants. They had been pouring into the city over previous years, and social panic about them and the cholera they might bring with them rose in lockstep. Prominent New Yorkers—progeny of earlier waves of immigrants themselves—demanded that the gates be slammed shut.

"Prevent further immigration to this country," Mayor Hugh Grant wrote to President Harrison in 1892, "until all fear of the introduction of cholera shall have disappeared." The country's newspaper of record agreed. "With the danger of cholera in question," *The New York Times* reported in a front-page article, "it is plain to see that the United States would be better off if ignorant Russian Jews and Hungarians were denied refuge here . . . These people are offensive enough at best; under the present circumstances they are a positive menace to the health of this country... Cholera, it must be remembered, originates in the homes of human riffraff."[29]

In 1893, amid rising hysteria about cholera-infected immigrants, New York City officials quarantined the *Normannia*, a vessel carrying immigrants from Hamburg, Germany, which had suffered cholera deaths en route. City officials wanted to hold the passengers at a hotel on Fire Island, but before they could disembark, armed mobs gathered on the docks and threatened to burn down the hotel. For two days, the mob jeered at the trapped passengers and barred them from leaving the vessel. Two regiments from the National Guard and the Naval Reserve had to be called in to allow them safe passage to land.[30]

⊙

Violent, cholera-fueled scapegoating in the nineteenth century intensified the pathogen's disruptive impact but probably didn't play much of a role in increasing cholera's death toll. Violence against physicians and immigrants surely reduced people's access to medical care, but given the state of medical treatments for cholera at the time—vast quantities of calomel, a mercury compound; tobacco smoke enemas; electric shock; and interventions such as plugging the rectum with beeswax, among others—that probably increased rather than decreased people's chances of surviving. The reverse is true now, because containment measures are actually effective. Today, when health-care workers and their containment measures are attacked, pathogens kill more people.[31]

During the 2014 Ebola epidemic in West Africa, health-care workers attempting to safely remove still-contagious corpses were chased, lied to, and assaulted. In Guinea's second-largest city, Nzérékoré, riots broke out when a team arrived to disinfect the local market. Near Guéckédou, villagers burned a bridge connecting their village to the main road to repel health-care workers. In another nearby village, a mob attacked a team of eight health-care workers, politicians, and journalists as they attempted to distribute information about Ebola. Two days later, their corpses—including three with slit throats—were found in the septic tank of the village's primary school. "We don't want them in there at all," a village chief in Guinea explained to *The New York Times*, referring to health-care workers. "They are the transporters of the virus in these communities."[32]

Commentators often explained West Africans' mistrust of Western medicine by pointing to their superstitious beliefs around disease transmission, but recent historical events in the affected countries probably played more of a role. People in Guinea, Liberia, and Sierra Leone had suffered more than two decades of human-rights violations and atrocities at the hands of the military before Ebola arrived, eroding public trust in authority figures. The fact that health-care workers, imbued with official authority, were mostly foreigners probably didn't help inspire confidence among the locals either.

In South Africa, the government itself attacked lifesaving containment measures: the antiretroviral medications that treated AIDS. At an international scientific conference on AIDS held in 1985, National Institutes of Health researchers had reported—on the basis of what turned out to be a faulty assay—that HIV had infected two-thirds of schoolchildren in Uganda, and up to one-half of the population of Kenya. The claim was grossly exaggerated, but the idea that the new virus originated in the "heart of darkness" struck a chord among Western journalists. Sensationalistic stories about HIV's impact in Africa became, as Kenyan president Daniel arap Moi put it, "a new form of hate campaign."[33] Outraged at Western scientists' and their news media's implication that Africans were to blame for HIV's spread, antiapartheid leaders such as South African president Thabo Mbeki dismissed the entire notion that HIV existed at all. AIDS, Mbeki said, was just a newfangled term for malnutrition and diseases of poverty.[34] For years Mbeki's government refused to provide AIDS medications to South African patients, and restricted the use of donated medications as well. (His administration touted the healing powers of lemon juice, beetroot, and garlic instead.) Between 2000 and 2005, more than three hundred thousand South African AIDS victims died prematurely for lack of effective treatment.[35]

In the United States, it was antagonism toward gay people and injecting drug users most at risk of HIV that thwarted early containment efforts. The Centers for Disease Control withheld funding for educational programs that included instructions on how to avoid HIV (through safe sex) that they deemed overly "explicit." The U.S. Senate prohibited financing for AIDS education materials if they "promoted" homosexuality. For more than two decades, the government banned federal funding for programs that provided injecting drug users with sterile syringes that would reduce their risk of contracting HIV for fear of sanctioning drug use.[36]

People with AIDS were fired from their jobs, deprived of insurance, health care, and other services, and subjected to violent assaults. In a 1992 survey, more than 20 percent of people with HIV or AIDS said they'd been physically victimized due to their HIV status. Haitians were similarly marginalized, after scientists noticed that clusters of Haitians had been infected with HIV. Many of these cases traced back to the ballooning epidemic in gay men, and the booming sex tourism trade that brought Western tourists into Haiti, but the idea that Haiti itself, with its unhygienic conditions and exotic Voudon rituals, had spread the virus captured the public imagination. "We suspect that this may be an epidemic Haitian virus," a National Cancer Institute physician told the press in 1982, "brought back to the homosexual population in the United States."[37]

"Haitians lost jobs, friends, homes and the freedom to emigrate," remembers the Haitian American writer Edwidge Danticat. "Children, including myself, were taunted or beaten in school by their peers. One child shot himself in a school cafeteria in shame." The Haitian tourism industry was destroyed.[38]

The arrival of West Nile virus in the United States provided another opportunity for people to blame despised—and wildly off the mark—scapegoats. Bioterrorism had long preoccupied certain sectors of the American political establishment, despite the fact that harnessing pathogens for warfare had rarely been attempted in modern times, and even then, mostly unsuccessfully. Members of the Japanese cult Aum Shinrikyo were known to have visited Zaire during an Ebola outbreak, but they apparently found that virus too difficult to weaponize. Besides that, and the poisoning of salad bars with salmonella in Oregon in 1981 by followers of the cult leader Bhagwan Shree Rajneesh, on the eve of West Nile virus's arrival, there'd been more angst about bioweapons than actual bioweapons themselves. (This was before the 2001 anthrax attacks on the United States, in which five were killed and seventeen sickened.)[39]

Nevertheless, when West Nile virus arrived in New York City in 1999, government officials were quick to suspect a bioterror attack at the hands of the hated Iraqi president Saddam Hussein.

Paltry evidence was brought to bear: the CDC had sent samples of West Nile virus to an Iraqi researcher back in 1985, and an Iraqi defector named Mikhael Ramadan had claimed that Hussein had weaponized the virus. Among other things, Ramadan said he'd worked as Saddam Hussein's double. "In 1997," Ramadan wrote in his 1999 memoir *In the Shadow of Saddam,* "on almost the last occasion we met, Saddam summoned me to his study. Seldom had I seen him so elated. Unlocking the top right-hand drawer of his desk, he produced a bulky, leather-bound dossier and read extracts from it," detailing the "SV1417 strain of the West Nile virus—capable of destroying 97 pc of all life in an urban environment."[40]

Even putting aside the gross exaggeration of West Nile virus's virulence—which is less than 1 percent, and which relies upon a complicated transmission sequence from birds to mosquitoes to humans, and does not actually spread directly from one person to another—Ramadan's account seemed fanciful. Even the tabloid newspaper that published an excerpt from the book, the *Daily Mail* of London, had to acknowledge that the book was possibly fraudulent, and its publisher admitted that it mostly just wanted to publish a good story. Nevertheless, *The New Yorker* published a long account by the writer Richard Preston detailing suspicions that Hussein had weaponized West Nile virus and unleashed it upon New York.

Bioweapons analysts at the CIA were "uneasy," Preston wrote. A top scientific adviser to the FBI told Preston that the fact that the West Nile virus outbreak seemed natural supported the idea that it was indeed a terrorist plot. "If I was planning a bioterror event," he explained, "I'd do things with subtle finesse, to make it look like a natural outbreak." Indeed, added Secretary of the Navy Richard Danzig, bioterrorism was "hard to prove." But it was also "equally hard to disprove."[41]

Even short epidemics, like the one SARS caused, have led to violent scapegoating. In 2003, hundreds of Canadians fell ill with SARS after a Toronto resident returned from Hong Kong infected with the virus. Two hospitals in Toronto had to be closed down, all nonessential medical services came to a halt, and thousands of people who had visited the hospitals voluntarily quarantined themselves for ten days. Spain and Australia issued warnings not to travel to the stricken city. In the panicked hysteria that followed, Asians of all stripes—whether they'd traveled abroad or not—found themselves singled out for social exclusion.[42]

Chinese Canadians were shunned on the subway.[43] "If you sneeze or cough," one remembered, "you could empty the train!" White Canadians pulled their jackets over their faces when passing Asians in hallways and wore masks in their offices if they had Asian coworkers. "As far as I'm concerned," one Asian Canadian overheard a coworker saying, "the whole community should be locked up." Families told their children not to play with Asian kids at school, employers withdrew job offers to Asian candidates, and landlords kicked Asian families out of their homes. Hate mail besieged organizations such as the Chinese Canadian National Council: "you people live like rats and eat like pigs and spread dirty, dirty, dirty disease around the world," one letter read. Losses to Chinese-owned businesses in Toronto reached up to 80 percent. "Asians were afraid to go anywhere," one Asian Canadian remembers.[44]

Epidemic-fueled violence has befallen other species, too. It made a certain amount of sense, in the beginning of the Lyme disease outbreak, to target deer. Early studies had found that the ticks that carry the disease fed on deer, and that on islands where deer were eradicated, tick populations fell. Plus, the nationwide deer population had zoomed from 250,000 in 1900 to 17 million by the mid-1990s. The animals rampaged through forests and destroyed suburban lawns and gardens.[45]

But follow-up research showed that deer had nothing to do with infecting ticks—the ticks pick up the bacteria that cause Lyme disease from rodents. Nevertheless, bloodthirst for the antlered animals rose.[46] Towns and counties across Connecticut, Massachusetts, New Jersey, and elsewhere expanded their deer-hunting seasons and opened up previously off-limits public lands to deer hunters. In Nantucket, orange-vested hunters from as far afield as Texas and Florida descended upon the island to stalk the animals. "Something has to be done," a Nantucket resident insisted. "People are going to die from this."[47] The History Channel created a reality television show to capture the rapidly expanding hunt, following camo-wearing deer hunters as they convinced the well-heeled residents of suburban Connecticut to allow them to shoot deer—"infamous for causing car accidents and spreading Lyme disease" as the show's website put it—on their property. (That series was titled, grievously, *Chasing Tail*.)[48]

The authoritarian government of Hosni Mubarak ordered a similar slaughter, this one of Egypt's three hundred thousand pigs, during the

H1N1 influenza pandemic of 2009. There was no evidence that pigs had played a role in spreading H1N1. The virus originated in pigs, which is why it was initially called "swine" flu, but it was a human pathogen: people caught it from each other. Egypt hadn't suffered even a single case of H1N1 flu at the time. Nevertheless, upon government orders, bulldozers and pickup trucks scooped up scores of swine. Some were killed with knives and clubs. "A large number of the pigs were herded into pits," *The Christian Science Monitor* reported, "and buried alive."

The bloodbath did little to quell H1N1's spread. It did, however, destroy the livelihood of the pigs' owners, the trash collectors called *zabaleen* of Egypt's embattled Christian minority.

In this case, scapegoating in reaction to one pathogen increased people's vulnerability to other ones. The pigs played an important role in protecting the public's health, by being used by the *zabaleen* to consume the organic portion of the household waste they collected door-to-door. In Cairo, their pigs processed 60 percent of the city's trash. Deprived of their pigs, the *zabaleen* stopped collecting trash altogether. When the government's attempt to replace them failed—the international waste collection companies the government hired expected Egyptians to pack their garbage into bins for periodic pickup, which they didn't like to do—trash accumulated on the streets, threatening Egyptians with filthborne contagions. "On any given day," wrote one visiting reporter, "a given neighborhood becomes a 'no man's land' of garbage." The slaughter of the pigs, said one community leader in Cairo, "was the stupidest thing they ever did . . . just one more example of poorly informed decision makers."

Ultimately, while the people of Egypt skirted the disease risks posed by their trash buildup, the Mubarak regime didn't fare as well—it fell during the Arab Spring revolution two years later.[49]

⊙

Epidemics are not the only social crises that have led to attacks on health-care workers and medical interventions. Vaccination campaigns have triggered similar waves of violent rejection and reprisals. But while the cause is different, the result is the same: efforts to contain pathogens are undermined, allowing epidemics to unfold.

Around the world, from the villages of northern Nigeria to the suburbs of Los Angeles, people have rejected vaccines and those who administer them, accusing them of a range of misdeeds, from undermining Islam to poisoning babies with chemicals. The WHO's campaign to eradicate polio, launched in 1998, is a good example. Rumors about the safety and purpose of the vaccine abound. In Nigeria, Muslim leaders said that the polio vaccine was contaminated with HIV and secretly meant to sterilize Muslims. The governor of Kano state halted the campaign for a year.[50] In North Waziristan, Pakistan, Taliban leaders claimed that the vaccine teams were a cover for an espionage campaign.[51] In Bihar and Uttar Pradesh in India, people alleged that the shot was contaminated with pig's blood and contraceptives.[52] And these suspicions often flared into violence. In northern Nigeria, vaccinators were assaulted and barred from entering homes. In Pakistan in 2012, militants began targeting vaccine workers, along with parents who agreed to have their children vaccinated. By 2014, they'd killed sixty-five vaccine workers.[53]

The reasons behind the violence are undoubtedly variable and rooted in local circumstances. But like societies gripped by epidemics of new disease, the Muslim societies that rejected Western-led vaccine campaigns were in the throes of a similarly existential crisis: a rising tide of anti-Muslim sentiment in the United States and Europe and the looming threat of military intervention. And Western vaccinators may have seemed like agents of destruction the same way that clinicians seemed the abettors of Ebola in the forests of Guinea and the Irish the conveyers of cholera in nineteenth-century New York. They had, in fact, been known to participate in coercive, secretive campaigns. During the 1970s campaign to eradicate smallpox, American vaccinators in South Asia had smashed down doors and held down crying women to administer the shots.[54] In the Philippines, people had been rounded up at gunpoint to be vaccinated against smallpox.[55] In 2011, the Central Intelligence Agency had used a sham hepatitis B vaccination campaign as a cover to collect information that led to the assassination of Al Qaeda leader Osama bin Laden in Pakistan.[56]

Whatever the reason, wherever vaccine refusals took hold, polio surged. And it spread. Nigeria's polioviruses spread into northeastern Ghana, Benin, Burkina Faso, Chad, Mali, Niger, and Togo. Poliovirus in India spread southward to Congo, triggering an outbreak among older, unvaccinated people. "We've got two hospitals with hundreds of paralyzed people and many dead," the WHO's Dr. Bruce Aylward told *The New York Times* in 2010. In just two weeks that year, Indian poliovirus paralyzed more than two hundred people in Congo.[57] Poliovirus from Pakistan migrated into China, which hadn't seen indigenous polio transmission since 1994, and, amid the wreckage of a brutal civil war, into Syria in 2013.[58] In 2014, the World Health Organization was forced to declare a global health emergency.[59]

Deep mistrust of vaccines and vaccinators has allowed once tamed pathogens to cause outbreaks in the United States and Europe as well. Despite the fact that vaccination played a decisive role in reducing cases of pertussis, measles, and chicken pox in the United States, when the government started requiring that children receive a battery of vaccinations before entering school in the 1980s, resistance to vaccines and mistrust of vaccinators mounted. Pop-music acts such as the Refusers railed against vaccination programs, as did celebrities such as the actors Jenny McCarthy and Jim Carrey. Thousands of websites assailing the risks of vaccination sprang up on the Internet.[60]

Vaccine refusals in the United States follow the same contours as they do elsewhere. The existential crisis that seems to fuel the mistrust in this case—roughly, the industrial contamination of nature—is more amorphous, but the vaccines and vaccinators targeted for reprisals are similarly imbued with malevolent power. One of the most popular antivaccine arguments is that the measles-mumps-rubella combination vaccine is endowed with the mysterious power to cause the poorly understood and increasingly common condition of autism. This claim is as exaggerated and conspiratorial as was the claim that doctors killed people with cholera to dissect their bodies in the nineteenth century, or that the polio vaccine is designed to sterilize Muslims. It's plainly contradicted by the facts. The 1998 research paper that alleged a link between the MMR vaccine and autism has been widely debunked and was withdrawn by the journal that published it. Plus, a 2013 study found that autism can be effectively detected in children at the age of six months, well before any would have been vaccinated against measles, negating any causal link between the two. The claim continues to make the rounds regardless.[61]

Another popular antivaccine claim is that drug companies push vaccines solely to make more money. This too is contrary to fact. Corporate influence on vaccine promotion is relatively slim. Indeed, drug companies have considered vaccines so unprofitable that during the 1990s and 2000s, many abandoned the vaccine business altogether. Between 1998 and 2005, nine vaccines required for routine childhood immunizations suffered chronic shortages as a result.[62]

But while vaccines neither cause autism nor drive the bottom lines of drug companies, they are the concentrated result of elaborate industrial processes. For people who fear industrial contamination, that's sufficient grounds to reject them. After all, vaccine skeptics who advise families to eschew vaccines don't object to the concept of immunization, in which bodies are exposed to a weakened pathogen to prophylactically build up immunity to it. The magazine *Mothering*, for example, which focuses on natural parenting techniques, suggests that in lieu of vaccination against chicken pox, families throw "pox parties," at which children infected with chicken pox purposely infect others. "Pass a whistle from the infected child to the other children at the party," the magazine advises. What they object to is not immunization but its delivery via the vaccine, a synthetic product of industrial processes that is injected directly into the body.[63]

The rage and frustration of the vaccine-promoting pediatricians and public-health experts on the receiving end of this mistrust is palpable. Nearly 40 percent of pediatricians surveyed by the American Academy of Pediatrics in 2005 claimed that they'd refuse to provide care to any family that rejected vaccines.[64] At a gathering of public-health experts in Atlanta in 2012, a speaker discussed the problems with the "antivaccinology" crowd. At one point, he broadcast an illustration of an MRI scan of the cartoon character Bart Simpson's head, his minuscule brain prominently highlighted. "I think this may be one of the antivaccinologists," he muttered theatrically, to titters from the audience. "No, I shouldn't say that!"[65]

As vaccine refusals have spread, the protections they provided against pathogens are fraying. Amid a rising tide of antivaccine suspicion, nineteen U.S. states allowed parents to exempt themselves from vaccinating their school-age children for "philosophical" reasons. Fourteen states, including California, Oregon, Maryland, and Pennsylvania, passed laws making it easier for parents to exempt their children from vaccination than to actually vaccinate them.[66] By 2011, more than 5 percent of kindergarteners in public schools in eight states had not been vaccinated.[67]

Seven percent of schoolkids in Marin County, one of the wealthiest counties in California, were unvaccinated for philosophical reasons. That's enough to undermine "herd immunity" against pathogens like measles, whereby pathogens are deprived of sufficient numbers of susceptible people to spread. Without herd immunity, pathogens can infect both unvaccinated people and those who can't be vaccinated, like infants.

Measles had been formally declared eliminated from the United States in 2000; by 2011, there'd been over a dozen new outbreaks, including one that began in late 2014 at the Disneyland theme park in California. Within two months, that outbreak had infected 140 people in seven states. (The governor of California eliminated personal and religious belief exemptions for vaccines a few months later.)[68]

Skepticism about vaccines, in particular the MMR, is even more widespread in Europe. In 2006, more than half of the French populace had not received the required two doses of measles vaccine.[69] Neither had 16 percent of the British populace in 2011.[70] An epidemic of measles started in Bulgaria in late 2009, spreading into Greece and ultimately to thirty-six countries across Europe. France and Britain were particularly hard hit.[71] By 2011, more than fourteen thousand people fell ill with measles in France. Across the Continent, measles sickened over thirty thousand.[72]

⊙

Is it possible to sever the connection between the amorphous fears inspired by epidemics and other social crises and the misplaced blame that follows? In the philosopher René Girard's conception, ending the cycle of fear and blame requires establishing the scapegoat's innocence, as in the New Testament story of Jesus's persecution, which ends with his resurrection.

Perhaps a modern form of accountability could perform a similar function today. In Haiti, human-rights lawyers are attempting to do just that, although not to establish innocence but to establish guilt. They are taking the United Nations to court to prove its complicity in the cholera epidemic. Soon after the epidemic exploded in Haiti, the Haitian human-rights lawyer Mario Joseph, in partnership with lawyers in the United States, collected fifteen thousand complaints from cholera-scarred Haitians. Since the Haitian government had granted the United Nations immunity from Haitian courts and the commission that the agency had said it would create in order to process claims against its troops had never been established, Joseph and his colleagues planned to sue the UN in courts in the United States and Europe, demanding an apology and reparations for the cholera epidemic.[73]

Joseph's office is in a grand house in Port-au-Prince, with a heavy door outfitted with brass adornments. Inside, it's dark and oppressively hot, with just a few ceiling fans whirring so slowly that I wondered, when I visited in 2013, why they'd turned them on at all. The windows are not glassed in, and the sounds of the permanent traffic jam on the road wafted in. Joseph, dark and round in a light blue short-sleeved dress shirt, railed about imperialist, racist interventions in Haiti, sweat beading on his forehead and neck.

"The Nepalese poured fecal matter in the river and a lot of people drink the river water," he said. "This is a calamity from this force of occupation!

"They need to pay people. They need to compensate people. And then they need to apologize for that! Because the United Nations never protected people from Nepali cholera! . . . Imagine if that happened in the United States, or in France, or Canada or England? What would happen? . . . I don't know, is it because we are Haitians? . . . Is it because we are black? I don't know! Because we are Haitian? I don't know!"[74]

Joseph's contention is that the United Nations understood the sanitary conditions in Haiti and thus should have taken adequate precautions to avoid introducing a dangerous new pathogen. It should have more rigorously screened its troops for signs of infection, for instance, and ensured that waste disposal practices on its bases were sanitary. By not doing so, it threw a match onto a gas-soaked pyre. It was not unfairly scapegoated: it was actually culpable. And holding the UN accountable in a court of law would show that.

There's little doubt that the cholera in Haiti really did come from Nepal. When scientists compared the genomes of the cholera vibrio circulating in Haiti with a sample of cholera vibrio from Nepal, they found a near perfect match, with only one or two base pairs distinguishing the two.[75] Still, I couldn't help but feel troubled by the underlying idea that even if people are judged to have infected others with pathogens, they should be held legally responsible. It's possible that neither sanitary waste management on the base nor rigorous health screening of the troops could have prevented cholera from Nepal from coming to Haiti. The person who introduced it was most likely a silent carrier, unaware of the undetectable infection within his body. Even if his waste had been treated on the UN base, it wouldn't have been almost anywhere else in Haiti. In that way, the Nepalese soldiers who brought cholera to Haiti were like any one of us. With the global biota and all its pathogens on the move, we're all potential carriers.

Channeling victims' fury in court is undoubtedly a lot more constructive than acting it out in the streets. But couldn't the judgment Joseph sought endow scapegoating with the force of law? Depending on who did the adjudicating, that group of "responsible" people could have included health-care workers in Guinea in 2014, gay people in the United States in the 1980s, and Irish immigrants in New York City in the 1830s. Even if those who introduced new pathogens were accurately pinpointed, as in Haiti, it's unclear how much of the blame they should be forced to shoulder. Epidemics are sparked by social conditions as much as they are by introductions. Whether it's deforestation and civil war in West Africa, the lack of sanitation and modern infrastructure in Haiti, or the crowding and filth of nineteenth-century New York City, without the right social conditions, epidemics of cholera and Ebola would have never occurred. Should health-care workers in West Africa, UN soldiers in Haiti, or Irish immigrants in nineteenth-century New York be held responsible for those, too?

As if to underline the selective nature of the blame game, in which UN soldiers are held accountable but local infrastructure weaknesses are not, in the middle of his rant, the dim fluorescent tubes in Joseph's office flickered and then clicked off. The various machines in the room— the printer, the computers, the lazily ineffective ceiling fan—all came to a halt. Soon the room was plunged into darkness. I leaned forward in my chair and looked around, preparing to make a move. But Joseph was unfazed. He knew that electricity in Port-au-Prince is spotty. "It will come back," he said calmly, and kept talking into the battery-powered recorder I'd set upon his massive desk.

☉

The ways in which new pathogens conspire to weaken our social ties and exploit our political divisions are wide-ranging and varied. But there's still one final way we can defang them. It is, perhaps, the most potent one of all. We can develop specific tools to destroy or arrest them with surgical precision, tools that can be effectively used by any individual with access to them, with no elaborate cooperative effort required.

Those tools, of course, are medicines.

The right cures make all the ways we spread pathogens among us moot. With the right cures, spillovers, filth, crowding, political corruption, and social conflict fail to spread pathogens. Epidemics and pandemics are stillborn, and their nonevents pass unnoticed. So long as there's a drugstore on the corner or a doctor willing to write a prescription, individuals can tame pathogens on their own.

First, though, those cures have to be developed.

SEVEN

THE CURE

Finding out how cholera spread and how to stop it was a matter of great urgency for every part of nineteenth-century society affected by the disease, but no sector more so than the medical community. The arrival of the deadly new disease charged the medical world like a bolt of lightning. By all accounts doctors and scientists worked intensely to unravel cholera's mysteries and save their stricken patients. They expounded on their ideas about the pathology and transmission of the disease in scores of papers, lectures, conferences, and essays. They developed a bevy of experimental treatments, theories as to how cholera spread, and interventions designed to arrest it.

And yet for decades, effective cures for cholera eluded them.

Their failure was not due to lack of technical capacity. The cure for cholera is almost comically simple. The vibrio does not destroy tissue, like, say, blood-cell-devouring malaria parasites or the lung-destroying tubercle bacilli that cause tuberculosis. It doesn't hijack our cells and turn them against us, like HIV. As deadly as cholera is, its tenure in the body is really more like a visit from an unpleasantly demanding guest than a murderous assailant. What kills is the dehydration the vibrio causes while replicating in the gut. That means that surviving cholera requires solely that we replenish the fluids it sucks dry. The cure for cholera is clean water, plus a smattering of simple electrolytes like salts. This elementary treatment reduces cholera mortality from 50 percent to less than 1 percent. Preventing cholera by separating human waste from drinking-water supplies was similarly well within the reach of nineteenth-century technology. The aqueducts and reservoirs of the ancients could have done it.[1]

The failure was not due to a lack of observations about the nature of cholera, either. Scientists and doctors had noted the association between cholera and dirty water since the earliest days of cholera's emergence in Europe. In Moscow, the epidemic ravaged the banks of the Moskva; in Warsaw, along the banks of the Vistula; in London, along the banks of the Thames. The French surgeon Jacques Mathieu Delpech noted in 1832 that the cholera in England spread from a central point to the periphery, and "that central point was the bank of the river." That same year, another French commentator observed that cholera spread from a fountain "full of putrid matter," and once abandoned, "there were no further cases of cholera."[2] In 1833, a medical professor had even published a map of the city of Lexington, Kentucky, which correlated the location of cholera deaths to the local topography and its filth. The same is true for the saltwater cure. It had been first proposed—and supported with solid evidence—in the 1830s.[3]

Nineteenth-century physicians had made the right observations and had the right technology to cure cholera. The problem was that the right observations and the right technology were beside the point.

⊙

In 1962, the physicist and philosopher of science Thomas Kuhn explained how the practice of science can paradoxically repress as much as it reveals. Scientists understand reality through the prism of what Kuhn called paradigms, theoretical constructs that explain why things function the way they do. Paradigms provide explanatory frameworks for scientific observations. They're like elaborate line drawings that scientists fill in with color and detail, reinforcing and enriching the paradigms as they do so. Evolution forms such a paradigm for modern biology; plate tectonics, for modern geology.

Hippocratic theory was nineteenth-century medicine's paradigm. According to Hippocratic principles, health and disease were the result of complex, idiosyncratic interplays among large, amorphous external factors, like meteorological conditions and local topography, and unique internal factors. Maintaining and restoring health was a matter of correcting the balance among these various factors.

These ideas had first been articulated by followers of the ancient Greek physician Hippocrates and had been handed down, essentially intact, for thousands of years. The fifth-century B.C. *Hippocratic Corpus*, a collection of sixty tomes on health and medicine, along with a ten-thousand-page elaboration on its ideas by the second-century A.D. physician Galen, had been standard issue in medical education since the sixth century. By A.D. 1200, professional licensure in Europe required the study of these works. Important English and French translations of Hippocratic and Galenic texts continued to appear throughout the nineteenth century.[4]

Kuhn believed that without such paradigms, science couldn't exist. The number of available facts and questions that can be asked is potentially infinite. Without a sense of why something might occur the way it does, he observed, there's no way for scientists to know which questions to ask or which facts to collect. There's no way to get to the "how" questions that underlie most scientific activity.

But as useful as paradigms are, they also create subversive dilemmas for scientists. Paradigms create expectations, and expectations limit scientists' perceptions. Psychologists have described two common cognitive snags that occur: "confirmation bias" and "change blindness." The problem of confirmation bias is that people selectively notice and remember only the subset of evidence that supports their expectations. They see what they expect to see. They also fail to notice anomalies that contradict their expectations, which is "change blindness." In one study of change blindness, experimenters purposely violated people's expectations by covertly switching one interviewer with a different person while the person being interviewed was momentarily distracted. Subjects assimilated the perceptual violation to such an extent that they didn't consciously register the change. It was as if it never happened at all.[5]

Confirmation bias and change blindness are two ways observations that violate expectations—or what Kuhn called anomalies, facts that subvert paradigms—are ignored. These two cognitive biases undoubtedly played a role in Hippocratic doctors' failure to notice when cholera didn't proceed according to Hippocratic principles. But Kuhn noted another cognitive dilemma, too. Sometimes, when people are forced to recognize anomalies, they still reject them.

Kuhn pointed to a 1949 study of cognitive dissonance in which subjects were asked to identify playing cards. Most were normal, but a few were anomalous, such as a red six of spades or a black four of hearts. When people were asked to identify such cards, "without apparent hesitation or puzzlement" they immediately identified them as normal cards, he noted. What the subjects saw was an anomalous red six of spades, but what they said they saw was an ordinary black six of spades or an ordinary red six of hearts. That's a form of confirmation bias. But what's interesting is what happened when they were shown the anomalous cards multiple times. Although they became increasingly aware that something was wrong with the cards, they were unclear what exactly it was. Some refused to accept the anomalies and became distressed. "I can't make the suit out, whatever it is it didn't even look like a card that time," one subject said. "I don't know what color it is now or whether it's a spade or a heart. I'm not even sure now what a spade looks like. My God!"[6]

The history of medicine is replete with examples of this phenomenon. When observations and treatments were unexpected or violated reigning paradigms—and no alternative explanation could be convincingly articulated—they were thrown out on theoretical grounds alone, no matter how well supported they were by evidence. In the seventeenth century, for example, a Dutch draper named Anton van Leeuwenhoek had handcrafted a microscope and discovered bacteria. He examined rainwater, lake water, canal water, and his own feces (among other things), and everywhere he looked he found microorganisms, which he called "animalcules." Further inquiries could have revealed the role these microbes played in human disease, but instead the study of the body through microscopy went underground for two centuries. The idea that tiny things shaped health and the body in some mechanical fashion violated the Hippocratic paradigm of health as a holistic enterprise. The seventeenth-century physician Thomas Sydenham, known as the "English Hippocrates," dismissed Leeuwenhoek's microscopic observations as irrelevant. His student, the doctor and philosopher John Locke, wrote that attempting to learn about disease by examining the body through microscopy was like trying to tell time by peering into the interior of a clock.[7]

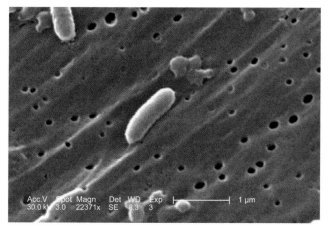

A scanning electron microscope image of *Vibrio cholerae* 01 (CDC / Janice Haney Carr, 2005)

A simulated flu pandemic on a map depicting locations and cases according to their temporal distance on the air travel network (Dirk Brockmann)

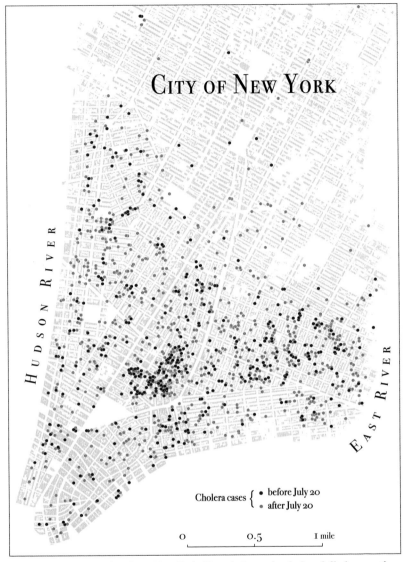

CITY OF NEW YORK

HUDSON RIVER

EAST RIVER

Cholera cases { • before July 20
               • after July 20

0   0.5   1 mile

The 1832 cholera outbreak in New York City. At its peak, cholera killed more than one hundred New Yorkers every day. (Sources: *The Cholera Bulletin, Conducted by an Association of Physicians*, vol. 1, nos. 1–24, 1832; base map adapted from *Map of the City of New York, 1854 . . . For D. T. Valentine's Manual 1854* using New York Public Library's Map Warper. Adapted by Philippe Rivière and Philippe Rekacewicz at Visionscarto.net from "Mapping Cholera" by the Pulitzer Center on Crisis Reporting at http://choleramap.pulitzercenter.org)

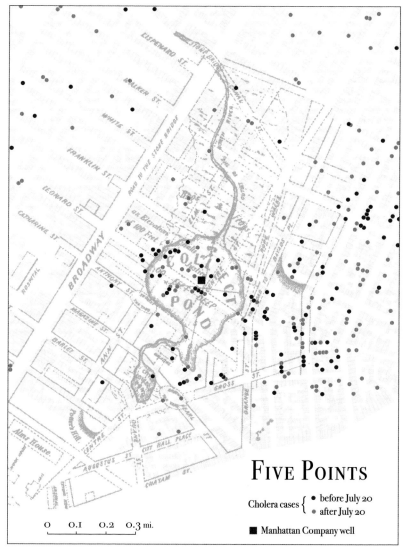

FIVE POINTS

Cholera cases { • before July 20
              { • after July 20

O   O.I   O.2   O.3 mi.

■ Manhattan Company well

The 1832 cholera outbreak in New York City. The Manhattan Company, now JPMorganChase, sank its well amid the privies and cesspools of the Five Points slum, atop the site of the Collect Pond, which had been filled in with garbage. This water was distributed to one-third of the city of New York. (Sources: *The Cholera Bulletin, Conducted by an Association of Physicians*, vol. 1, nos. 1–24, 1832; base maps adapted from *Map of the City of New York, 1854 . . . For D. T. Valentine's Manual 1854* and John Hutchings, *Origin of Steam Navigation, a View of Collect Pond and Its Vicinity in the City of New York in 1793*, 1846, using New York Public Library's Map Warper. Adapted by Philippe Rivière and Philippe Rekacewicz at Visionscarto.net from "Mapping Cholera" by the Pulitzer Center on Crisis Reporting at http://choleramap.pulitzercenter.org)

The sparkling interior of Medanta Hospital, New Delhi. The hospital caters to some of the hundreds of thousands of medical tourists who visit India for surgeries and other medical procedures. By 2012, medical tourism had spread the antibiotic-resistant superbug New Delhi metallo-beta-lactamase-1 (NDM-1) to twenty-nine countries around the world. (Sonia Shah)

A drain pipe from Medanta Hospital grounds exiting into a trash-strewn gutter. Pathogens such as NDM-1 have been found in New Delhi's drinking water and surface waters. (Sonia Shah)

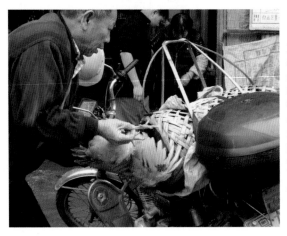

Packing up birds at the Jiangcun poultry market, Guangzhou, Guangdong province, China. H5N1 influenza emerged in Guangdong in 1996 thanks in part to transmission opportunities provided by giant poultry farms. (Sonia Shah)

An illegal pig farming colony, Laocun, Gongming, Shenzen, Guangdong. The farmers, their families, and the pigs live together in the low shacks. Epidemiologists speculate that pigs coinfected with human and avian influenzas may have allowed H5N1 influenza to acquire the ability to infect humans. (Sonia Shah)

The waterfront at Cité Soleil, Port-au-Prince, Haiti. In 2006, less than 20 percent of the population of Haiti had access to toilets or latrines. Human waste dumped on empty lots blocked by garbage threatened drinking-water supplies, much to cholera's advantage. (Sean Roubens Jean Sacra)

The sole tap with regularly running water in Belle-Anse, Haiti. A pig wallows in the muck. (Sonia Shah)

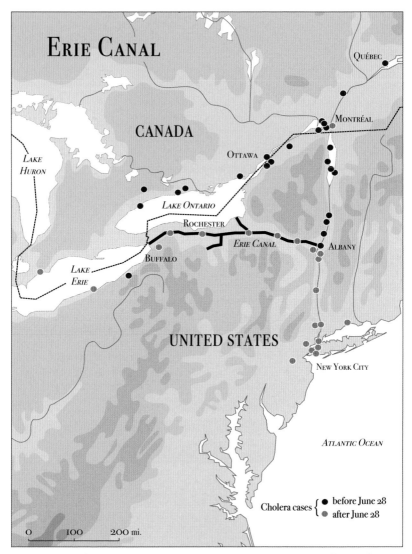

Cholera cases along the Erie Canal, 1832. Government physicians collected this data in 1832 but denied that the canal or the Hudson River had anything to do with the spread of cholera. Neither was quarantined. (Sources: Data compiled by Ashleigh Tuite from Lewis Beck, *Report on Cholera. Transactions of the Medical Society of the State of New York*, 1832. Adapted by Philippe Rivière and Philippe Rekacewicz at Visionscarto.net from "Mapping Cholera" by the Pulitzer Center on Crisis Reporting at http://choleramap.pulitzercenter.org)

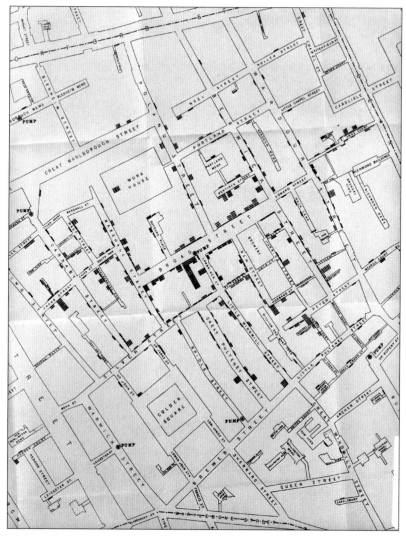

The anesthetist John Snow's map of the 1854 cholera outbreak in Soho, London, in relation to the Broad Street pump. Snow proved that cholera was transmitted in contaminated water, but the medical establishment didn't accept his findings until the 1890s. (Wellcome Library, London)

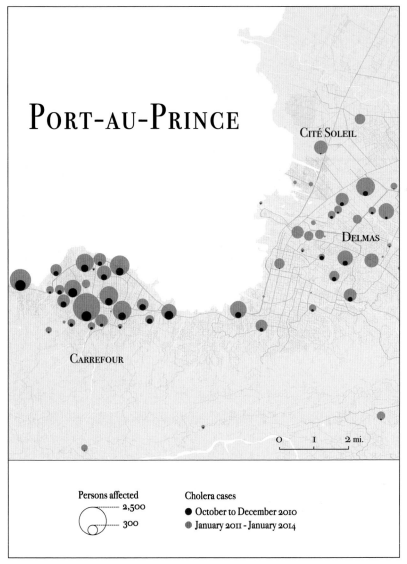

PORT-AU-PRINCE

CITÉ SOLEIL

DELMAS

CARREFOUR

O   I   2 mi.

Persons affected
.... 2,500
.... 300

Cholera cases
● October to December 2010
● January 2011 - January 2014

The cholera epidemic in Port-au-Prince, after Nepalese soldiers working as UN peace-keepers introduced cholera into Haiti in 2010. Within a year, there were more cholera victims in Haiti than everywhere else in the world combined. (Source: Weekly tallies of cholera cases treated at MSF clinics provided by Médecins sans Frontières, 2014; base map from OpenStreetMap. Adapted by Philippe Rivière and Philippe Rekacewicz at Visionscarto.net from "Mapping Cholera" by the Pulitzer Center on Crisis Reporting at http://choleramap.pulitzercenter.org)

Similarly, in the eighteenth century, the shipboard doctor James Lind had discovered that lemon juice cured scurvy, a condition of vitamin C deficiency, by the unorthodox method of comparing the outcomes of different groups of sailors who were given different treatments. He's lauded today for conducting the first clinical trial. But at the time, because he couldn't support his theory as to why lemon juice worked (that is, according to Lind, because acidic lemons supposedly broke through the pores blocked by damp air), his findings were dismissed. Medical experts recommended ineffectual vinegar rather than lemon.[8]

This is just what happened to cholera's cures in the nineteenth century. The scientists who discovered cholera's cures were not as fully indoctrinated in the paradigms of Hippocratic medicine as the elite physicians atop the medical establishment. They were outsiders. William Stevens, for example, was a lowly physician who had practiced medicine in the Virgin Islands and was unknown among Britain's medical elites in London. So was the Scottish physician William O'Shaughnessy. Both advocated for cholera's lifesaving cure of salty water in the 1830s. Stevens thought it would help correct cholera patients' dark-colored blood. (He had noticed that salt reddened the blood of his tropical fever patients.) O'Shaughnessy recommended "the injection into the veins of tepid water, holding a solution of the normal salts of the blood," as *The Lancet* reported, not only to fix the color of the blood but also to restore the body's lost fluids and salts.[9] In one of the most convincing demonstrations of the therapy's effectiveness, in 1832 Stevens administered salty fluids to more than two hundred cholera sufferers at a London prison and lost less than 4 percent of his patients.[10]

But the logic of the cure—replenishing the fluids lost through vomiting and diarrhea—violated Hippocratic paradigms. According to Hippocratic principles, epidemic diseases like cholera spread through foul-smelling gases called "miasmas" that poisoned those who inhaled them. That's why cholera patients experienced dramatic vomiting and diarrhea: their bodies were attempting to get rid of the miasmatic poison. Counteracting those symptoms with salt water or anything else was as philosophically wrongheaded as ripping a scab is today.

And so medical experts lambasted the salt water advocates. Experts who visited the prison to review Stevens's results dismissed them out of hand, claiming that the patients he treated never had cholera to begin with. They defined a cholera victim as someone on his or her deathbed, in the throes of what they called "collapse." As none of Stevens's patients lay in such a state, by definition they could not have cholera. (The outlandish notion that they might actually have recovered was inconceivable.) "There certainly was not a single case which from any symptoms witnessed by me I could point out as a case of cholera," asserted one investigator. One young woman, a "very troublesome perverse character," another pointed out, was just "simulating" cholera.

Journal editors who reviewed Stevens's work concluded that he was a charlatan. "We turn from the whole affair with mingled sensations of pity and disgust," the editors of *The Medico-Chirurgical Review* wrote. "The best that can be hoped for the 'saline treatment' and its authors is, that both may be speedily forgotten."[11] "However it might be with pigs and herrings," punned one commentator in 1844, "salting the patients was not always the same as curing him." Saline therapy, another agreed in 1874, had "proved ineffective."[12]

Evidence that showed that cholera spread in dirty water, not miasmas, was similarly dismissed and suppressed. The nineteenth-century London anesthetist John Snow was particularly well placed to grasp the shortcomings of the miasma theory as applied to cholera. Snow had been knocking himself unconscious with various gases—ether, chloroform, and benzene, among others—for years, studying their effects on his body in search of the perfect anesthetic to administer to his patients.[13] As an expert on the behavior of gases, he knew that if victims contracted cholera by inhaling a gas, as the medical establishment held, the disease would affect the respiratory system, including the lungs, much like a deep breath of acrid smoke. And yet, it didn't. Instead, cholera plagued the digestive system.[14]

For Snow that could mean only one thing: cholera must be something that victims swallowed.[15] Snow collected potent evidence in support of his theory. During a cholera outbreak in Soho in 1854, Snow went door-to-door interviewing local residents. He found, by plotting his results on a map, that nearly 60 percent of the residents who drew water from a popular drinking-water pump, on Broad Street, had sickened with cholera, compared to only 7 percent of those who didn't. He even found out how the water got contaminated. With the help of a local priest named Henry Whitehead, he located a woman named Mrs. Lewis, who lived at 40 Broad Street near the well. They learned she had washed her cholera-infected baby's diapers in a partially blocked cesspool situated less than three feet away from the pump's shaft.

Finally, Snow correlated death rates with London residents' drinking-water sources. Some of the city's water companies distributed contaminated downstream water, while others tapped upstream waters, out of reach of the city's sewers. In 1849, when both a water company called Lambeth and another called Southwark and Vauxhall drew water from the contaminated lower Thames, Snow found, both of the districts they served suffered equal numbers of cholera deaths. But after the Lambeth Company moved its intake pipes farther upstream, the death rate among its customers dropped to one-eighth that of the customers of the Southwark and Vauxhall Company.[16]

Snow had mustered a masterful case that dirty water, not miasmas, caused cholera. The trouble was that his findings undermined the fundamental tenet in miasmatic theory. It was as if he'd told a bunch of biologists that he had found life on the moon. Accepting such a seditious claim required disavowing principles that had governed medicine and medical practice for centuries.

The medical establishment responded much as the subjects of the anomalous card experiment did. They tried to call a red six of spades a black six of spades, assimilating and submerging Snow's claims into the edifice of miasmatism. A committee was convened by the board of health to consider Snow's findings. However agitated they may have felt, they did not reject them outright, most likely because although Snow was an outsider in terms of his expertise, he was an eminent figure among medical elites, who'd administered chloroform to the laboring queen and served as the president of the London Medical Society. And so, in their voluminous report—more than 300 pages, including a 352-page appendix with 98 tables, 8 figures, and 32 color plates—the committee agreed that cholera could indeed spread in water. But, they reported, that did not mean that Hippocratic principles were mistaken. Cholera could grow in *either* air or water, they said, and of the two modes, air played the decisive role. "It is not easy to say which of these media may have been the chief scene of poisonous fermentation," the committee wrote. But on the whole, "it seems impossible to doubt that the influences . . . belong less to the water than to air."[17]

This convoluted yet unmistakable dismissal might have discouraged some scientists, but not Snow. He continued to insist that miasmatic theory was wrong. Finally, the medical establishment was forced to denounce him outright. Not long after the committee issued its report, Snow testified in a parliamentary hearing against a board of health bill cracking down on industries that emit miasmas. The parliamentarians railed against his antimiasma stance. "Are the Committee to understand, taking the case of bone-boilers, that no matter how offensive to the sense of smell the effluvia that comes from bone-boiling establishments may be, yet you consider that it is not prejudicial to the health of the inhabitants of the district?" parliamentarians demanded. He assented. "You mean to say," they followed up, "that the fact of breathing air which is tainted by decomposing matter, either animal or vegetable, will not be highly prejudicial to health?"

When Snow refused to budge, an edge of hysteria crept into the parliamentarians' line of questioning. "Do you not know that the effect of breathing such tainted air is to produce violent sickness at the time? . . . Have you never known blood poisoned by inhaling putrid matter? . . . Do you not know that the effect of a very strong offensive smell is to produce vomiting? . . . Do you dispute the fact that putrid fever and typhus fever hang about places where there are open sewers?"[18]

The Lancet, Britain's leading medical journal, reviewing Snow's unflinching testimony against the bill, accused him of betraying the public's health. "Why is it . . . that Dr. Snow is singular in his opinion? Has he any fact to show in proof?" the journal's editors fumed. "No! But he has a theory, to the effect that animal matters are only injurious when swallowed! . . . The well whence Dr. Snow draws all sanitary truth is the main sewer . . . In riding his hobby very hard, he has fallen down through a gully-hole and has never since been able to get out again."[19]

Rebuking Snow, Parliament passed the health board's anti-stinky-fumes bill in July 1855. Other than convincing the local parish in Soho to remove the handle of the contaminated Broad Street pump (which probably didn't have much effect on the cholera outbreak, since it had collapsed by then anyway), Snow's findings on cholera left hardly a ripple on miasmatism's imperturbable surface.

In 1858, Snow abandoned the study of cholera to write his magnum opus on anesthesia, *On Chloroform and Other Anaesthetics*. On June 10, as he wrote the final sentences, a paralyzing stroke catapulted him off his chair and onto the floor.[20] He died six days later. *The Lancet*, still miffed at Snow's disavowal of miasmatism, printed an abbreviated obituary, which pointedly didn't mention his incendiary work on cholera.[21]

⊙

With cholera's subversive cures squashed, nineteenth-century doctors continued to administer their Hippocratic treatments and interventions. These kept long-held principles comfortably intact.

They also made cholera much worse.

Nineteenth-century treatments for cholera increased its death toll from 50 to 70 percent.[22] Since they considered cholera patients' vomiting and diarrhea therapeutic, doctors treated patients with compounds that intensified the very symptoms that were killing them. They administered the toxic mercury compound mercurous chloride, or "calomel," which induced vomiting and diarrhea.[23] (Calomel, the eighteenth-century American physician and surgeon general Benjamin Rush wrote, was a "safe and nearly universal medicine.")[24] Physicians literally poisoned their patients with it, considering treatment complete only when the patient salivated excessively, his mouth turned brown, and his breath started to smell metallic—all signs that physicians today would recognize as mercury toxicity.[25]

They drained their cholera patients' blood. "Bloodletting" had long been a cure-all therapy, enthusiastically endorsed by Galen himself. Each spring, people would line up to go to the doctor to be bled, just as they had since medieval times.[26] The thinking behind the practice was that the removal of blood would restore the four "humors" whose interactions within the body and the environment shaped health status. Physicians considered it especially salutary for cholera patients, because it got rid of their strangely dark, thick blood (which Stevens had tried to rectify and which is now understood to be a sign of dehydration.[27] "All the practitioners who have seen much of, and have written upon, this disease agree upon one point," remarked Dr. George Taylor in *The Lancet* in

1831, "the immense advantage of bleeding at the commencement of the disease."[28]

Most egregiously, they promoted the dumping of human waste into the drinking-water supply. According to miasmatic theory, the flush toilet, or "water closet," improved human health by rapidly ridding human habitations of bad smells. People in London started installing them in the late eighteenth century. Because they considered the smells dangerous but the actual excreta harmless, they cared little about where the waste was dumped, so long as it was distant enough to save their noses from the odors. Thus they sent their waste through the sewers into the most convenient dump site, the Thames River, which ran through the city. The more excreta they deposited in the river, the safer they felt. Preventing cholera and other diseases required, as *The Times* put it, ensuring that there was "no filth in the sewers," which carried toilet waste, and "all in the river." The city's sewer commissioners proudly noted the huge volume of human waste the city's toilets efficiently deposited into the river: twenty-nine thousand cubic yards in the spring of 1848, and eighty thousand cubic yards by the winter of 1849.[29] They even discerned a correlation between mortality and the filthiness of the river. London's death rate had declined, *The Times* reported in 1858, as "the Thames has grown fouler."[30]

In fact, it was just the opposite, since London relied upon the Thames for its drinking water. Twice a day, when the tide in the North Sea rose, reversing the Thames's downstream current, the plume of filth they dumped into the river flowed as far as fifty-five miles upstream into the intake pipes of the city's drinking-water companies. Nevertheless, under the powerful influence of miasmatism, when cholera struck, Londoners believed it was not because too many flush toilets dumped human waste into the river but because *too few* did. Flush toilet sales, in the years after the 1832 epidemic in London, enjoyed "rapid and remarkable" growth, according to an 1857 report. Flush toilet sales enjoyed another spike after an 1848 outbreak of cholera. So many Londoners installed flush toilets over the 1850s that the city's water use nearly doubled between 1850 and 1856.[31]

Hippocratic medicine, in other words, was a boon to *Vibrio cholerae*. It had probably played a similarly helpful role to other pathogens, too. Over its long tenure, historians judge Hippocratic medicine to have harmed more people than it helped. Nevertheless, as a system of thought it was remarkably resilient. It explained health and disease more effectively than the earlier ideas—that health and disease were the work of the gods—that it replaced. As Thomas Kuhn observes, "To be accepted as a paradigm, a theory must seem better than its competitors, but it need not, and in fact never does, explain all the facts with which it can be confronted."[32] That was certainly true of Hippocratic medicine. And once it was accepted as a paradigm, it took on its own momentum, helped along by the cognitive biases of its practitioners. Hippocratic principles themselves were "wonderfully versatile as an explanatory system," as the historian Roy Porter has written. Its conception of the body as ruled by four bodily fluids (blood, phlegm, black bile, and yellow bile) could be correlated with all manner of external phenomena, such as the four seasons, the four stages of human development (infancy, youth, adulthood, old age), and the four elements (fire, air, water, earth). Physicians had been making correlations and adding nuance to Hippocratic principles for centuries, enriching it with layers of depth and meaning.[33]

And it was able to maintain an illusion of success. The kind of group-to-group comparisons that could have revealed the uselessness of its treatments were rarely undertaken, because Hippocratic medicine considered patients as "unique as snowflakes," as one epidemiologist put it. There was no point in grouping them together and comparing their outcomes. If, in general, patients treated with mercury fared poorly compared to patients treated with other remedies, Hippocratic physicians would never have known.[34] Furthermore, although some Hippocratic treatments were harmful, many were probably just useless. These could have seemed beneficial, thanks to placebo effects, by which medical interventions work simply because patients believe they will. (Experts suggest that the placebo effect may be responsible for a third of the apparent effectiveness of modern medicine.)[35] Thus, while "for 2,400 years patients have believed that doctors were doing them good," the historian David Wootton writes, "for 2,300 years they were wrong."[36]

☉

In a serendipitous twist, Parliament decided to start resewering London, a project that ultimately ended cholera transmission in the city, the very year that John Snow died. The fact that the resewering commenced so soon after Snow's death has often led casual observers to conclude that the British medical establishment had accepted Snow's insights. On the contrary, the resewering had nothing to do with John Snow and every-thing to do with miasmas.

The medical establishment had been agitating for a new sewer system for years before Snow made his claims about cholera. The social reformer Edwin Chadwick had advocated for new sewers in his hugely popular report, "On an Inquiry into the Sanitary Condition of the Labouring Population of Great Britain," in 1842.[37] The reason Chadwick and others wanted to resewer the city was to get rid of miasmas, more specifically, the gases that arose from the sewers. High tides regularly prevented the sewers from dumping their loads in the river, forcing the sewage inside the pipes to back up into the city. Miasmatists didn't consider that in itself a problem, except that the backups allowed stinky sewer gases to escape and waft into unsuspecting Londoners' nostrils, which they considered a serious public-health hazard. Stories of men who'd entered the sewers and been instantly suffocated by sewer gases abounded.[38]

But while miasmatic experts agreed on the value of resewering the city, they disagreed about just how to do it. Chadwick thought a whole new network of sewers should be built, to flush the gases (and the waste) into distant waterways and farms. Others, such as the surgeon and chem-ist Goldsworthy Gurney, thought sewers could simply be ventilated of their gases. The gases could then be passed through a steam bath and burned to render them odorless, and thus harmless. With the city un-able to resolve the dispute, neither plan moved forward until a conflu-ence of strange weather events made the stench of the sewers impossible to ignore.[39]

A heat wave hit the city in the summer of 1858. This followed a period of drought that had lowered the water level in the Thames River, expos-ing the thick layer of excreta that blanketed its banks. As the mercury climbed to over 100° Fahrenheit, a powerful stench started to emanate from the dried-up, waste-covered banks of the river.[40]

Newspapers dubbed the odor "the Great Stink."

For people who believed that smell caused disease, the Great Stink augured catastrophe. Panic set in. "What is to become of us

Londoners in the year of grace 1858?" the *British Medical Journal* asked. "Is this vast metropolis to be devastated by the plague?" "Everyone is crying out that something must be done!" the *Medical Times and Review* wrote. "We know that the stench from the mud-banks and from the water itself is so great," wrote *The Lancet*, "that strong and healthy become faint, and even vomit, and that it produces fever."[41] People were being "struck down with the stench," added the *Journal of Public Health and Sanitary Review*, "and of all kinds of fatal diseases, upspringing on the river's banks." "It stinks," the *City Press* wrote, "and who so once inhales the stink can never forget it and can count himself lucky if he live to remember it."[42]

The Thames had never been as stinky. More important, never before had the most powerful people in the city been as exposed to its odors. Thanks to a recent reconstruction, the Palace of Westminster where members of Parliament congregated enjoyed more than eight hundred feet of riverfront.[43] And the elaborate ventilation system, which had been devised to protect their chambers from miasmas—and which would have sealed their chambers from the odors of the river—had been dismantled. That ventilation system had drawn in fresh air from the top of one of the palace's three-hundred-foot towers, filtered the air through damp sheets and sprayed it with water, then forced it into the chambers of Parliament through thousands of tiny holes drilled in the floor. To minimize drafts, a rough horsehair carpet blanketed the floor holes. The system then extracted the "used" air through raised panels built into a glass ceiling. The windows in Parliament's chambers, which opened into London's stinky air, were never opened.[44]

But when parliamentarians complained of bouts of dizziness in 1852, the ventilation system was blamed and its designer (whom *The Times* called an "aerial Guy Fawkes") dismissed.[45] Goldsworthy Gurney—the man who believed that London's sewers could continue to dump their loads into the river, so long as the sewer gases were combusted—took over the job of protecting Parliament from miasmas. One of his first acts was to pry open the windows, allowing the most powerful men in London to become intimately acquainted with the sights, sounds, and odors emanating from the river below.

Thus, as the Great Stink rose from the river, it drifted into the open riverfront windows of Parliament's chambers.[46] Legislators fled the library and their meeting rooms, for fear of "being destroyed by the present pestilential condition of the River Thames," as *The Times* noted. The leader of the house escaped while holding a handkerchief over his nose. The lords abandoned their committee rooms. There was worried talk of moving Parliament elsewhere. The stink interrupted the Court of Queen's Bench, as physicians warned that "it would be dangerous to the lives of the jurymen, counsel and witnesses" if anyone remained in the reeking chambers. With the Great Stink enveloping the palace, by the end of June, Gurney sent notice to his bosses that he could no longer be responsible for the health of the House of Parliament.[47]

In fear of a plague, parliamentarians introduced legislation to speed the process of resewering the city. Their options were now more clear. With Gurney in disgrace, they adopted an alternative plan, that of the engineer Joseph Bazalgette. He proposed to build a system of inter-cepting sewers, which would divert all of London's gases—and its waste—farther downstream of the city. And so, what John Snow's map could not accomplish, Goldsworthy Gurney's open windows did. By 1875, the new sewers had been built, Bazalgette had been knighted, the Thames was sewage-free, and cholera had been banished from London forever.

It all happened while the medical establishment continued to repu-diate the idea that waste-contaminated water transmitted cholera.[48]

The city of New York happened to clean up its water supply around the same time, also while rejecting the idea that dirty water spread cholera. There, the precipitating factor was the demand, from the city's brewers, for better-tasting water to brew their beers. For fifty years—and two explosive cholera epidemics—the Manhattan Company had dis-tributed contaminated groundwater to New Yorkers, even as residents pled for better-tasting water and a more abundant supply for fire extin-guishing and street cleaning. But when the brewers added their voices to the outcry about the dirty water, which put their beers at a competi-tive disadvantage, the business-friendly city council finally resolved to rectify the problem.[49] By then, the Bronx River could no longer feasibly supply sufficient clean water, so the city tapped the distant Croton River, requiring a forty-two-mile-long aqueduct.[50]

Croton water started to flow into New York City in 1842.[51] Few people signed up at first. But after cholera struck in 1849, they lined up by the thousands. By 1850, the water department's annual revenue nearly doubled compared to the previous year. The voluminous Croton water allowed the city to flush its once stagnant sewers, too, and it started expanding its network of sewers in the 1850s. By 1865, the city had built two hundred miles of sewer pipes to carry its wastewater out to the rivers. The last cholera outbreak in New York City, in 1866, took fewer than six hundred lives. And then the disease vanished from the city for good.[52]

Neither city had any awareness that it had implemented strategies based on Snow's antimiasmatist insights. Londoners attributed the cessation of cholera to the diversion of stinky sewer gases; in New York, cholera's demise was attributed to the street-cleaning efforts of the board of health. "If we had had no Health Board," a newspaper editor noted at the time, "we would probably have had a great deal more cholera."[53]

That may not have mattered much for those two cities, which solved their cholera problems despite miasmatism, but it made a big difference elsewhere. While New Yorkers and Londoners enjoyed cholera-free lives of clean water and sanitation, their revolutionary way of life was as likely to ignite social changes elsewhere as a wet match. Cutting-edge late nineteenth-century architecture in Naples, Italy, for example, focused on elevating buildings above low-lying miasmas, leaving them bereft of easy access to clean drinking water.[54] The ultramodern city of Hamburg, Germany, in the late nineteenth century distributed unfiltered river water contaminated with sewage to its residents with proud efficiency.[55]

Across much of the European continent, the teachings of the German chemist Max von Pettenkofer, who'd convinced attendees of the International Sanitary Conference in Constantinople in 1866 to disavow John Snow's drinking-water theory, prevailed. For Pettenkofer, poisonous clouds caused cholera. This belief had all kinds of untoward ramifications, not least in his advocacy, during cholera outbreaks, of flight. Pettenkofer said that when the poisonous cholera clouds formed, "rapid evacuation" was always a "salutary measure." When cholera broke out in Provence, France, in 1884, Italian authorities went so far as to distribute free rail tickets and charter steamships to rapidly evacuate Italian immigrants—and their cholera—out of infected France. Back in Italy, they proceeded to seed a new outbreak in Naples.[56]

Until miasmatism was renounced by the medical establishment, and replaced by a new paradigm that could integrate cholera's cures into its explanatory framework, such physician-approved practices that abetted cholera would continue.

⊙

That new paradigm arrived in the late nineteenth century. "Germ theory" posited the idea that microbes, not miasmas, cause contagions. The theory rested on a spate of discoveries. Microscopy had finally come back into fashion, allowing scientists to revisit the microbial world first spied by Leeuwenhoek two centuries earlier. And then, by conducting experiments on animals, they determined the specific role these microbes played in animal diseases. The French chemist Louis Pasteur discovered the microbial culprit behind a disease of silkworms in 1870; the German microbiologist Robert Koch discovered that *Bacillus anthracis* caused anthrax in 1876.[57] These findings were still incendiary to miasmatists, but they were fundamentally different from the ones they'd rejected in the past. They didn't arrive on the scene sporadically, seemingly out of nowhere, but with an increasingly steady regularity. And they were encased in a powerful explanatory framework. Germ theory, along with accounting for the nature of contagions, provided a radically new way to think about health and illness more generally. Rather than being the result of complex disequilibria involving amorphous external and internal factors, poor health was now discernible at the microscopic level.

In 1884, Koch made a splash at a conference on cholera in Berlin by announcing he'd discovered the microbe responsible for causing cholera: *Vibrio cholerae*. (In fact, Koch wasn't the first to spy the bacteria—an Italian doctor named Filippo Pacini had isolated what he called a "choleraic microbe" in 1854.) And Koch had developed a method of proving that the bacteria caused the disease. His method, known as "Koch's postulates" and used up until the 1950s, involved a three-step proof. First, he extracted the offending microbe from a patient ill with the disease. Second, he grew the microbe in the lab on a nutrient-laden petri dish. Third, he administered the lab-reared microbe to a healthy individual. If that person fell ill with the disease in question, the microbe was proved to be the culprit.

But Koch couldn't accomplish the proof for *Vibrio cholerae* and cholera (infecting experimental animals with *Vibrio cholerae* is notoriously difficult).[58]

"Koch's discovery alters nothing," leading miasmatists such as Pettenkofer scoffed, "and, as is well known, was not unexpected by me." Other experts called Koch's discovery "an unfortunate fiasco." An 1885 British medical mission (the head of which considered Pettenkofer "the greatest living authority on the etiology of cholera") reported that the vibrio Koch had discovered had nothing to do with cholera at all.[59]

To prove that the bacteria didn't cause cholera, Pettenkofer and his allies devised a daring demonstration. Pettenkofer procured from a patient dying of cholera a vial of stool, teeming with hundreds of millions of vibrios, and drank it.[60] The fluid went down "like the purest water," he proclaimed. Twenty-seven other prominent scientists, including Pettenkofer's assistant, did the same. A popular Paris magazine covered their shenanigans with a drawing of a man eating feces while shitting a bouquet of violets. The caption read: "Dr. N. consumes a cholera-ridden feces orally; five minutes later, he produces a bouquet of violets . . . at the other end."[61] Although both Pettenkofer and his assistant came down with cholera-like diarrhea, the assistant suffering bouts every hour for two days, all of the cholera drinkers survived, which Pettenkofer considered a successful repudiation of Koch's germ theory.[62]

The stand-off between miasmatism and germ theory continued for several more years. Then an 1897 outbreak of cholera in Hamburg sealed miasmatism's fate. According to miasmatic theory, the city's western suburb of Altona, which like Hamburg lay along the banks of the Elbe River, should have fallen prey to the miasmas that caused cholera in Hamburg as well. And yet it didn't. It was impossible for experts to deny the reason why: Altona filtered its drinking water, while Hamburg did not. Strikingly, none of the 345 residents of an apartment block called Hamburger Hof—within the political boundaries of Hamburg but receiving water from Altona's filtered supply—fell ill at all.[63]

With this stark vindication of Koch's claims (and the long-dead Snow's), miasmatism's last advocates were forced to surrender. Hippocratic medicine, after a two-thousand-year-long reign, had been knocked off its throne. In 1901, Pettenkofer shot himself in the head and died. A few years later, Koch received the Nobel Prize in Medicine and Physiology. The germ theory revolution was complete.[64]

With the end of miasmatism, cholera's terrifying tenure in North America and Europe entered its closing years. The gains against cholera in London and New York spread. Municipalities across the industrial world improved their drinking water through filtration and other techniques. After 1909, when liquid chlorine became available, municipalities started chlorine disinfection.[65] The few waterborne pathogens that survived twentieth-century water treatment and filtration grew milder.[66]

The germ theory revolution improved treatments for cholera as well. The Anglo-Indian pathologist Leonard Rogers proved that salty fluids cut cholera mortality by a third in the early 1900s, after which the once mocked injections of saline gained popularity.[67] Scientists steadily refined rehydration therapy over the course of the twentieth century. Today, a saline solution mixed with a bit of lactate, potassium, and calcium is an antidote to cholera as effective as a shot of insulin to a patient in a diabetic coma. Oral rehydration therapy, by simply and quickly curing cholera and other diarrheal diseases, is considered one of the most important medical advances of the twentieth century.[68]

That's not all. There are vaccines against cholera, too, which aim to replicate the protective immunity enjoyed by cholera survivors by delivering whole killed cells and subunits of cholera toxin. Although nobody yet knows how that immunity works, products such as Shanchol, a cheap oral vaccine licensed in 2009, and the traveler's vaccine Dukoral are nevertheless 60–90 percent effective, at least for a few years, making them useful additions to the anticholera arsenal.[69] (Since they require multiple doses and take several weeks to become effective, the WHO recommends that they be used in conjunction with other cholera prevention measures; as of this writing, neither is available in the United States.) Even simpler methods pioneered by the microbiologist Rita Colwell and her colleagues, such as filtering untreated water through a few layers of sari cloth, catches 90 percent of the vibrio bacteria in contaminated water, reducing cholera infections by 50 percent.

Medicine finally figured out how to cure cholera. It just hadn't done so fast enough to deliver humanity from nearly a century of cholera pandemics.[70]

⊙

Today, when new pathogens emerge, there's no decades-long delay in figuring out how they spread. Modern biomedicine quickly identifies new pathogens' modes of transmission. That HIV spread through sexual contact, and SARS through aerosols was obvious from the very first clusters of cases that occurred. The rapidity of medicine's ability to unravel modes of transmission allows preventive strategies to be promptly devised, like condoms in the case of HIV, or face masks in the case of SARS, or safe burials in the case of Ebola.[71] (Of course, that doesn't mean those strategies can or will be implemented—HIV infected nearly 75 million people worldwide by 2014 despite the medical establishment's insights into how to prevent it by practicing safe sex.)

But we still can't rely on modern medicine to save us from the threat posed by new pathogens.

For one thing, even when scientists devise new cures, we are not necessarily able to produce them at the right scale and at the right time. Drug development is slow and constrained by the economic concerns of the for-profit pharmaceutical industry. If the market for a new drug is modest, it doesn't matter how big the public-health need for it is, or how solid the scientific evidence supporting its effectiveness: that drug is unlikely to get to market. There are precious few drugs developed for diseases, like malaria and Ebola, that selectively afflict the poor. Malaria sickens hundreds of millions of people every year, but since most of those victims have less than $1 a year to spend on health care, the market for new malaria drugs is vanishingly small. Today, the most cutting-edge drugs available for the disease are based on a botanical compound called artemisinin, a two-thousand-year-old Chinese remedy. Ebola affects far fewer people than malaria but poses a much more alarming public-health threat. As of 2014, there were no drugs or vaccines available for Ebola, either. "'Big Pharma,'" as a headline in London's *The Independent* put it in 2014, "failed deadly virus' victims."[72] That means that pathogens that prey upon the untreated poor can amplify and spread into broader populations.

Another problem in counting on medicine to save us from emerging pathogens has to do with how new paradigms in medicine have replaced the old. Modern medicine may have no equivalent to the *Hippocratic Corpus* that is studied with Talmudic intensity, but its guiding philosophy is equally pervasive. Modern biomedicine's fundamental approach to solving complex problems is to reduce them to their smallest and simplest components. In its estimation, heart disease is a problem of cholesterol molecules in the blood; human consciousness is a chemical reaction in the brain. Each minuscule component of the complex phenomena of health and disease is studied by specialized experts, usually in isolation.[73]

When my doctors learned I had contracted MRSA, for example, they didn't ponder the landscape, or my home environment, or my immune status, or the animals that lived in my house, or my diet. They targeted the bug and only the bug. MRSA existed on one side of an invisible divide, I was on the other, gun in hand.

Modern medicine's reductionist approach is the exact opposite of Hippocratic medicine's approach, which was fundamentally holistic and interdisciplinary, and called upon a range of expertise to elucidate disease processes, from engineering and geography to architecture and the law.[74] That is no coincidence. Germ theory and the reductionist approach it represented was a revolutionary new paradigm for medicine. Revolutionary new paradigms generally don't accommodate old ones, subsuming their principles and approaches. They destroy the old ideas and purge the ranks of loyalists.

The limits of reductionism became clear to me during my struggle with MRSA. One of the worst abscesses I developed started during a summer holiday, which over the course of a week turned from a stinging pinprick to a slow-motion volcano of pus and blood that debilitated my leg to the extent that I could not comfortably walk or drive. I doused my body daily with hospital-issue disinfectant. I changed the dressing twice a day, and all my clothes, too. When the long-worn bandages started to irritate the skin under them, which had turned itchy and angrily red, I rushed to the store to search for something better, finding new "nonirritating" bandages to replace the old bandages I'd bought, now revealed as the "irritating" kind. (Who knew?)

My overriding fear was that the MRSA-filled pus would seep into one of the new fissures opening up under the bandages and tape that secured the dressing, allowing the pathogen to establish an even deeper foothold. The words of the microbiologist echoed in my mind: *He could have lost a leg.*

The MRSA basket in my bathroom expanded to become a MRSA shelf; my medical supplies to battle the microbe grew to include a messy crowd of boxes of sterile pads, tape, antibiotic cream, and drawing salve, something I'd read about online somewhere.

The battle continued like this for years. The abscesses kept coming back, in the same mysterious places. And every time, I redoubled my antimicrobial efforts, with more boiling of cloths, more wiping of counters, more drugs, more sprays, more bleach baths to rout out the intruder.

Finally, by year three of MRSA, I stopped fighting. Not for any good reason. I just got tired. One day a bump appeared, and though I noticed it, I could not bring myself to deal with it. I did not scratch or squeeze or apply ointment or heat or bleach. And, incredibly, it went away on its own. I didn't revel in the triumph. I figured it was just a one-time thing. But it happened again and again. It was as if once I stopped fighting it, it lost gusto for the fight, too. The abscesses seemed to get smaller and less noticeable. If I was patient enough, in time, without prodding, without any intervention at all, they'd quietly go away on their own.

I have no idea why this happened. Had my immune system figured out how to quell MRSA's appetites? Had some other strain of *Staphylococcus aureus* in my body repressed its growth? Was it my diet or exercise regime that undermined its ability to spread? Or perhaps it had nothing to do with me at all. Perhaps my symptoms had been the result of the anti-MRSA treatments themselves, or something in my environment. Whatever happened, I suspect it had to do with more than just the microbe my doctors and I had surgically focused our ire upon. There was some kind of Hippocratic interplay going on, between internal factors and possibly external ones, too.

Modern medicine, singularly focused on the microscopic, is poorly suited to grasp such interactions. And yet most of our new crop of pathogens similarly cross disciplinary boundaries. Pathogens in animals, studied by veterinarians, spill over into people, studied by physicians. But because the two fields rarely interact, the crossovers escape detection. Ebola virus afflicted chimps and apes before the 2014 epidemic in West Africa. Could the human outbreak have been caught earlier had doctors and vets been collaborating all along? West Nile virus killed crows and birds for a month before the human outbreak in New

York City occurred. In that case, it was a veterinary pathologist at the Bronx Zoo who finally linked the two outbreaks and pinpointed the virus as West Nile.[75] It's not just the experts who've drifted apart; patients consider the two fields separate and unrelated, too. Less than a quarter of HIV patients ask their vets about the health risks posed to them by their pets. These risks include salmonella (carried by turtles and other reptiles), MRSA (carried by dogs and cats), and, before the importation of African rodents was banned in 2003, monkeypox from pet prairie dogs.[76]

Biomedical experts rarely collaborate with social scientists. In one survey of biomedical experts, around half admitted to being "unreceptive" toward social sciences. Most of the others expressed ambivalence.[77] (They mostly objected to the messiness of social science research, compared to the controlled experimentation that medicine relies on.) And so when new pathogens cause outbreaks, biomedical causes and solutions are immediately sought, while social and political factors—like John Snow's sidelined discoveries about contaminated water in the nineteenth century—are treated as minor contributory factors. When West Nile virus broke out in New York City, the containment strategy revolved primarily around attacking the biomedical causes of the disease: the insect vector that carries the virus. The nonbiomedical factors, such as the loss of diversity among bird species, went unaddressed.

Social and economic factors that contributed to a 2009 outbreak of dengue in Florida were similarly ignored. In 2008, South Florida had suffered a rash of foreclosures, which had allowed mosquitoes to breed in abandoned swimming pools and gardens out of reach of mosquito inspectors and homeowners, leading to an explosion of mosquitoes. The following year, dengue broke out for the first time in seventy years, hitting particularly hard in Key West, the epicenter of the foreclosure crisis. A CDC study found that 5 percent of the population there harbored antibodies to dengue. But biomedicine's reductionist noncollaborative approach would hardly have led anyone to consider addressing the housing crisis as part of an appropriate response to the outbreak.[78]

⊙

Since the mid-twentieth century, biomedicine has been rightly celebrated for its powerful ability to render lifesaving cures. But its limits, which have already started to show, will only become more apparent in the coming years. Some of the external disruptions that are now eclipsing microscopic mechanisms as drivers of new disease are more amorphous, wide-ranging, and unpredictable than the world has ever seen.

EIGHT

THE REVENGE OF THE SEA

If there's any single historical development that actuated all of the various ways human activity contributes to pandemics, it's the harnessing of fossil fuels such as coal, oil, and gas. Before the discovery of coal and oil, civilization drew its energy primarily from wood fires and human labor. To acquire more energy, society had to expend a nearly equal amount of energy, whether it was in the form of felling trees or feeding slaves. There wasn't much of an energy surplus, which limited the size of human populations and their expansion across the globe as well as the frequency and scale of pandemics.

The discovery of rich veins of coal and buried reservoirs of oil liberated society from those thermodynamic constraints. The best fossil fuels can provide one hundred times more energy than their extraction requires.[1] The energy surpluses they unleashed allowed civilization to expand at a previously unimaginable clip. Each manifestation of fossil-fueled power—whether in the form of increased agricultural yields made possible through petrofertilizers or in the speed and scale of trade and transport—contributed to the emergence and spread of pathogens. Petrofertilizers doubled agricultural yields, feeding the growth of populations and their crowding into cities. Coal powered the steamboats that carried cholera across the oceans and the canal-building machines that ferried it into the interior of continents. Oil powered the machines that cut down the forests and the airplanes that dispersed the viruses they once concealed across the globe.

But in addition to fueling the population growth, urbanization, and mobility that contribute to pandemics, the global bonfire of fossil fuels will heighten the likelihood of pandemics on its own, in a way that is likely to be even more consequential than all of its contributing factors put together. The voraciousness and speed with which we consumed fossil fuels—one hundred thousand times faster than they could form underground—assured that. It was like eating a lifetime's supply of food at a single meal. The energy in fossil fuels, which derived from their carbon, had accumulated underground for millions of years. By digging it up and burning it, we released all of that ancient carbon into the atmosphere in a matter of decades, an outburst that would alter the climate and all the creatures that lived within its confines for generations.

By the mid-twentieth century, the concentration of carbon dioxide in the atmosphere had increased by more than 40 percent compared to preindustrial levels. Hanging in the atmosphere like a blanket, the excess carbon steadily warmed the air below, gently heating the surface waters of the ocean. Every decade, the temperature of the seas' surface waters rose by a little over one-tenth of a degree. Newly warmed waters, sinking to the depths and splashing into planetwide flows, altered the ocean's constitution in subtle but transformative ways, like a shot of vodka in a glass of tomato juice. Currents, fueled by the temperature gradient between cool waters moving over warm ones, were transformed. Rainfall patterns around the globe shifted, as the growing cloud of vapor that wafted above the warmer seas swelled by 5 percent. The warming waters, expanding as they heated up, lapped ever higher on coastlines and beaches, inundating freshwater habitats with salt water. By 2012, in some parts of the world, the sea had risen eight inches above 1960 levels.[2]

As the seas changed, so did cholera's fortunes.

⊙

For most of the twentieth century, cholera's connection to the sea was unknown. The sea itself was considered a static, unchanging place, a vast expanse "of eternal calm," as the environmental writer Rachel Carson put it, "its black recesses undisturbed by any movement of water more active than a slowly creeping current."[3] Scientists held plankton, those microscopic creatures floating in the ocean, in similar regard. Plankton, they believed, blanketed the languid seas uniformly, like a layer of dust on a mantelpiece. And it had nothing to do with cholera. Cholera vibrio, according to conventional wisdom, lived on land, traveling from one person's gut to another via contaminated drinking water.

A mild-mannered zoologist named Alister Hardy thought otherwise. He devised a simple but ingenious little machine that would revolutionize scientific understanding of plankton. It was a long, continuously moving roller that, when towed behind a boat, unfurled a band of silk that captured samples of plankton. Since it didn't need much by way of expert handling to install and didn't take up much room, ships of all kinds could drag the spools of silk behind them, collecting billions of samples of plankton for scientists to analyze. (The first to do so was the *Discovery*, the ship that had earlier carried the explorers Robert Falcon Scott and Ernest Shackleton to the Antarctic in 1901.)[4]

As Hardy's machine was dragged across the sea, cholera's life underwater slowly came into focus. The microbiologist Rita Colwell made the unexpected discovery of *Vibrio cholerae* in the waters of the Chesapeake Bay in 1976.[5] Although she couldn't culture the vibrio in the lab—that is, get it to produce colonies on little plastic dishes of agar (which microbiologists considered the "gold standard" for identifying bacteria)—by exposing her samples to fluorescent antibodies that would bind to the bacterium, she could see the vibrio glowing. She knew they were there.[6]

So she continued to sample coastal waters for *Vibrio cholerae*. And everywhere she looked, she found it: in ponds, rivers, lakes, and seawater from five continents. Ultimately, Colwell and other scientists discovered more than two hundred serogroups of *Vibrio cholerae* that lived in the sea, including types that produced cholera toxin and types that did not. They discovered how they lived, too: in conjunction with zooplankton, especially copepods.[7]

Meanwhile, Hardy's machine, now known as the Continuous Plankton Recorder, had compiled one of the most extensive and longest-running records of marine life in the world. By the beginning of the twenty-first century, it had been dragged across more than 5 million nautical miles of the North Atlantic. The spools of silk revealed that far from being dustlike and uniform, plankton are as exquisitely sensitive to their environment as the tiny trembling hairs on a spider's leg. They responded to subtle signals in the sea and air—the temperature of the surface waters, the northern boundary of the currents of the Gulf Stream—that operated over thousands of miles of ocean.[8]

And the changing conditions in the North Atlantic had clearly affected them. Starting in 1948, their biomass had plummeted sixfold. Decades later, the plankton had returned, but they weren't the same. Warm-water plankton species had shifted six hundred miles, reacting to the increasingly warm surface of the sea by moving north at the rate of fourteen miles a year.[9]

These shifts, in turn, dictated the fate of the cholera vibrio that lived in and on plankton. The insights revealed by Hardy's machine, coupled with Colwell's research, had pioneered a new understanding of the role of environmental microorganisms in shaping life on Earth. Whatever was going on with cholera had as much to do with what occurred under the waves as with what transpired in the lives and habits of people on land.

⊙

After causing nearly a century of continuous pandemics, cholera had seemingly disappeared in 1926, receding into its ancestral homeland in the Bay of Bengal. "As a world scourge," the historian William H. McNeill wrote in his landmark 1977 book on the role of infectious disease in history, cholera had been "effectively defeated." Its demise exemplified "an unusually tidy paradigm" of "triumphant containment."[10]

In fact, cholera hadn't disappeared in 1926, exactly. The particular strain that had ravaged the world in six pandemics—now known as "classical 01" *Vibrio cholerae*—had died out. But before vanishing it had spawned a sneaky little descendant, one particularly well suited to exploiting new opportunities presented by the changing sea it lived in. This new kind of cholera vibrio could thrive in rivers, estuaries, lakes, and ponds at least three times longer than classical 01.[11] It was a peculiarly resilient creature, able to withstand an onslaught of antibiotics, too.[12]

Although public-health experts did not recognize it as a pandemic-worthy pathogen until the 1970s, the descendant had first been spotted in 1904 at El Tor quarantine station on the west coast of the Sinai Peninsula, where it had been extracted from the corpses of six pilgrims to Mecca who'd died of diarrhea. At the time, compared to the cholera raging across the globe caused by classical 01, this new vibrio seemed insignificant. Investigators decided it wasn't a cholera vibrio at all but just some other kind of generic, forgettable vibrio. They named it, simply, after the place they'd found it: El Tor vibrio.[13] And then the medical establishment basically forgot about it.

El Tor vibrios resurfaced in 1937, causing outbreaks on the Spermonde archipelago, a series of isolated, low-lying coral atolls off the coast of South Sulawesi, Indonesia. Still, they escaped international attention.[14] The vibrio killed 65 percent of those it infected, but since the outbreaks did not spread beyond remote Sulawesi, the global health authorities at the World Health Organization didn't consider them to be caused by cholera. The disease that El Tor vibrios caused was just some kind of "peculiarity," the agency said, "conditioned by local circumstances." They called it "paracholera" and decided that nothing much should be done to contain it. "Quarantine, strict isolation of sick persons and their contacts, disinfection and mass immunization," the WHO reported, "are not justified."[15]

This would turn out to be a significant missed opportunity. For as environmental conditions in Spermonde changed, so did the nature of El Tor's outbreaks. Over the following years, increasingly voluminous rains, more powerful storms, and rising seas pummeled Sulawesi. Every year, rainfall increased by two to three inches. Storms became so ferocious that even the experienced fishers of the islands regularly lost their boats at sea. Rising seas permanently contaminated wells with salt water.[16]

In 1961, El Tor "paracholera" dramatically extended its reach, striking out of Sulawesi into other parts of Indonesia as well as the Philippines, Malaysia, and Thailand. By the summer, El Tor broke out in Guangdong province in south China, where it killed between thirty thousand and fifty thousand people, Western commentators estimated. According to their reports, whole villages had been razed. From there it seeped into Hong Kong and eventually into South Asia, cholera's heartland.[17] Since it was still traveling incognito as paracholera, not real cholera, the international rules regarding quarantines and notifications that pertained to cholera didn't apply.[18]

El Tor arrived in Africa, where the disease had never been seen before, in 1971.[19] It struck a huge gathering that took place on the banks of Lake Chad, a freshwater lake bordered by Chad, Cameroon, Niger, and Nigeria, for the circumcision ceremony of an important sheikh. Over eight hundred sickened and more than one hundred perished in a matter of weeks. The shallow, warm, plankton-choked body of water proved an excellent home for the environmentally resilient El Tor vibrio. Thanks to a frenzy of dam building, irrigation diversions, and land clearings along its coast, the lake was well on its way to drying up. By 2000, the lake, which had once covered over ten thousand square miles, extended over fewer than six hundred square miles, with a depth of less than five feet. Regular, deadly outbreaks across the Lake Chad basin followed, year after year.[20]

Finally, the WHO admitted that paracholera, a supposedly mild form of cholera-like illness peculiar to specific remote locations, didn't exist. El Tor *was* cholera, in all its terror-inducing, virulent glory. After four decades of "triumphant containment," cholera was back. A seventh pandemic had begun.[21]

⊙

In 1990, cholera arrived in South America, where it had not been seen since 1895.

Once again, its arrival coincided with a peculiar climatic phenomenon, in this case, El Niño Southern Oscillation, or ENSO. ENSO occurred every two to seven years, usually around December, which is why locals had named it after the baby Jesus, whose birthday they celebrated around the same time. It began when the trade winds failed, freeing the warm waters around Indonesia to drift eastward.[22] That warm patch of water unleashed rain clouds into the air above it, which acted like a boulder dropped into a stream, interrupting a range of other climatic patterns around the globe, leading to dry winters in the northwestern United States, heavier rain in East Africa, and more bushfires in north Australia.[23]

When El Niño's warm patch of water collided into the west coast of Peru in late 1990, it changed the composition of local plankton as well as the currents around the coast: local zooplankton populations crashed as equatorial zooplankton populations rushed in. The prevailing current, which flowed north up the coast, reversed direction.[24] Any cholera vibrio in those waters would have been made more abundant, more resilient, and more deadly by its warmth. Warm water helped cholera vibrio produce the toxins that desiccated their human victims. And it helped the bacteria attach to plankton, allowing it to survive for longer and in harsher conditions.[25] (Clinging to the egg sac or lining the gut of a copepod, vibrio concentrations could reach up to five thousand times higher than when free-living, and the bacteria could persist for more than a year.)[26]

Not long afterward, people who lived along a six-hundred-mile stretch of Peruvian coastline started falling ill with El Tor cholera.[27] Public-health authorities urged Peruvians to refrain from interacting with the newly deadly waters along the coast. Police arrested street vendors selling fish, including those selling the national dish, ceviche, because it consisted of raw fish marinated in citrus.[28]

Nevertheless, by the spring of 1991, cholera had sickened seventy-two thousand Peruvians and had started to spread across the continent. Rivers carried cholera into Ecuador, Colombia, and Brazil, and to the border of the United States. Cholera-rich surf washed onto the beaches of Los Angeles, compelling the hit television show *Baywatch* to consider fleeing north of the city. Cargo ships, their holds full of cholera-rich ballast water, dumped cholera into Mobile Bay, Alabama, leading to the closure of local oyster beds. An Aerolineas Argentinas flight carried cholera from Buenos Aires to Los Angeles, in the cholera-laced shrimp salad it fed to its passengers, sickening several dozen and killing one. Cocaine smugglers brought it into the remote southern Mexico villages where they stashed their secret airstrips.[29]

By 1993, nearly a million had sickened across Latin America, and some nine thousand were dead. Only Uruguay and the Caribbean had escaped El Tor cholera's fury. But not for long.[30]

⊙

With El Tor cholera increasingly abundant in the environment, by 1994 it had picked up a new trick, possibly by acquiring genes from its predecessor: the ability to secrete the same kinds of killer toxins that classical 01 had in the nineteenth century. Now, in addition to being more resilient in the environment and tougher against antibiotics than its predecessor, El Tor would be as efficient a killer as classical 01 had been.[31]

In Africa and Asia, El Tor's new toxin-producing ways ratcheted the death toll upward. Between 2001 and 2006, the proportion of cases in which it caused life-threatening dehydration rose from 30 to nearly

80 percent.[32] In 2007, "altered" El Tor had become the dominant cholera strain in South Asia, including in Nepal. Three years later, a group of soldiers hired by the United Nations escaped a local outbreak and boarded planes en route to the mountainous, earthquake-damaged island of Hispaniola, altered El Tor burning in their bellies.[33]

Haiti was a ticking bomb for a cholera explosion, but not just because of its history of strife, poverty, and deficient sanitation. Environmental conditions conspired to welcome cholera vibrio, too.

Up until 2010, pathogen-rich Haiti had been strangely immune to cholera. Cholera had first arrived in the Caribbean in 1833, with an outbreak in Cuba. But while the disease spread over the region, including the Dominican Republic, which occupied the eastern two-thirds of Hispaniola, there is no historical record of the disease showing up in Haiti. The Haitian historian Thomas Madiou speculated in the late 1850s that something peculiar about Haiti's geography protected it—"some emanations of our soil that don't allow cholera toxins to survive," he wrote, or "some condition of our atmosphere." If so, that protection vanished in the wake of the magnitude 7.0 earthquake that hit in January 2010.[34] Silt and limestone washed into the rivers, creating the kind of high-nutrient, alkaline conditions that vibrios love. The traumatized population was even more malnourished and poorly housed than they had been before. "There were very unnatural conditions in Haiti because of the earthquake," says the cholera expert Anwar Huq. "Nutrients came out of the ground. The ecology changed."[35] Ten months later, cholera finally annexed Haiti.[36]

The seventh pandemic, fueled by the most wily, resilient, and deadly cholera strain the world had ever seen, would be the longest and most widespread cholera pandemic ever. It continues to this day.[37]

⊙

Rita Colwell's research on *Vibrio cholerae*'s secret history in the sea catapulted her to the highest echelons of scientific research, including a six-year stint as director of the National Science Foundation. By the time the seventh pandemic enveloped Haiti, she was seventy-six years old. The impact of marine vibrios on humans had never been more apparent. Climate change had rendered the seas increasingly warm, and vibrio infections rose around the world, not just in Haiti. In the warming North Sea and Baltic Sea, vibrio infections soared.[38] In the United States between 2006 and 2008, vibrio infections grew by 43 percent. Pathogenic vibrio had spread into places they'd never been problematic before, like Alaska and Chile and Iceland, infesting shellfish and threatening those who eat it.[39]

I met Colwell in her office at the far end of the sprawling University of Maryland campus in College Park, where she is a distinguished professor, in the fall of 2011. (She is also a distinguished professor at Johns Hopkins University, and the chair of two microbial-detection companies.) She is well aware of the paradigm shift her work had triggered. "Thirty years ago," she said, "we were ridiculed to even say that the bacterium existed in the environment. But now it is in textbooks, the evidence is so overwhelming! It is understood!" Even after all these years, she still sounds surprised.

But Colwell is not done shaking up the scientific establishment. It is no peculiarity of cholera that the environment shapes its dynamics, she said. As the climate changes, the environment will play an equally salient role in the dynamics of other, novel infectious diseases. Inside cholera's story was a new explanatory framework for understanding emerging diseases, one in which the environment—biological, social, political, and economic—is both the source and the driver. This insight, Colwell said, has such far-reaching implications that it's tantamount to a scientific revolution, on the order of the paradigm shift from Hippocratic medicine to germ theory. She calls it the Cholera Paradigm.[40]

⊙

Determining just how the changing climate will influence infectious diseases is not straightforward. Odd weather combinations shape outbreaks of infectious disease in unpredictable ways. It was a cold snap in winter that led wild mute swans to alter their migration patterns in 2006 and ferry H5N1 into more than twenty European countries.[41] It was a mild winter in 1999 that allowed mosquitoes to breed in the sewers of New York City all season, followed by a summer drought, which forced thirsty birds to congregate in crowded watering holes, leading to the city's first West Nile virus outbreak.[42]

Clearly environmental conditions shaped these outbreaks, but could anyone have predicted just how? Consider an environmentally sensitive pathogen like *Plasmodium falciparum*, which causes mosquitoborne malaria. More rainfall can lead to more malaria—since it creates puddles and ponds that malaria-carrying mosquitoes breed in—or less, since runoff and floodwaters wash mosquito eggs away. Similarly, droughts can lead to more malaria by turning rivers into stagnant, mosquito-friendly ponds—or less malaria, since dry weather desiccates mosquitoes' bodies.

Still, certain correlations between weather and infectious disease are clear. The heaviest rainfalls (those in the top 20 percent) preceded 68 percent of the outbreaks of waterborne diseases that occurred between 1948 and 1994 in the United States.[43] Cases of West Nile virus rise by 33 percent after heavy rainfall.[44] And scientists agree that warmer temperatures will expand the range for the kinds of creatures that bring us diseases: bats, mosquitoes, and ticks among them.[45] It's already started. In Costa Rica, certain bat species have moved into higher-than-normal elevations, and in North America they have expanded their wintering ranges to the north.[46] The mosquito carrier of yellow fever and dengue, *Aedes aegypti*, long restricted to the southeastern Gulf states, popped up in California in 2013.[47] The Asian tiger mosquito, *Aedes albopictus*, spread northward and into higher latitudes in Italy.[48] Ticks have expanded northward and into higher latitudes in northern Europe and the eastern United States.[49]

Warmer weather makes life easier for these disease vectors. It can speed up their life cycles, too. Bark beetles destroy the tissues of trees by laying their eggs in tunnels under their bark. In warmer weather, the beetles can switch from a two-year life cycle to a one-year life cycle. Since the late 1990s, the beetles have been attacking increasingly younger trees and a wider range of species of trees, decimating nearly 30 billion conifers from Alaska to Mexico. In some states, like Wyoming and Colorado, one hundred thousand lodgepole pine trees decimated by beetles fall every single day.[50] A climatically sped-up life cycle may be one reason why. Other pathogens can similarly accelerate their cycles. Malaria parasites can cut days off their developmental cycle when the ambient temperature rises. That makes it more likely that they'll develop into infective forms within the short life span of their mosquito carriers.

Thus as our climate shifts toward warmer temperatures, hotter seas, and more volatile precipitation, cholera and its progeny will likely benefit. Simply by altering the distribution of pathogens, climate shifts will increase the burden of disease, as populations are exposed to new pathogens to which they lack immunity.

But that's what we can predict about the pathogens with which we are already acquainted. What about the pathogens we haven't yet met? According to the microbiologist Arturo Casadevall, the rising of Earth's ambient temperature could unleash whole new kingdoms of them.

⊙

We live in a world saturated with fungi. We ingest dozens of fungal spores with every inhale, and stomp through the world on ground that teems with fungi.

Fungi can be potent pathogens. Unlike viruses, which require living cells to survive, fungi can persist even after all their hosts are dead because they feed on dead and decayed organic material. They also can survive independently in the environment in the form of highly durable spores.[51]

Fungi are major pathogens of plants, as any backyard gardener knows. Some, such as *Phythophthora infestans*, which caused famine-producing potato blight, changed the course of human history. Others, such as the bat scourge *Pseudogymnoascus destructans* and the amphibian-plaguing chytrid fungus, have brought whole species to the brink of extinction.[52]

And yet while pathogenic bacteria and viruses regularly plague humans, aside from the odd yeast infection or case of athlete's foot, we suffer from very few fungal pathogens. This may be a function of our warm-bloodedness, Casadevall says. Unlike the reptiles, plants, and insects that regularly fall prey to fungal pathogens, mammals keep their blood at a scorching temperature—more than 35° above Earth's average ambient temperature of 61° Fahrenheit—regardless of the weather around us. Most fungi, adapted to environmental temperatures, can't take the heat of our blood and perish in our oven-like bodies.

Heat is such an effective antidote for infection that reptiles try to do it, too, producing "artificial fevers" by sunning themselves and thus raising their internal temperatures when suffering infections. Similarly, scientists have shown that warming frogs' bodies to 98° Fahrenheit cures them of chytrid fungus infection.

Warm-blooded mammals' superior defense against fungal pathogens,

Casadevall speculates, may explain the mystery of how mammals came to dominate over reptiles after the extinction of the dinosaurs. The cold-blooded lifestyle is far more efficient than ours. Warm-bloodedness requires mammals to consume ten times more daily calories than we'd have to if we were cold-blooded.[53] "You mammals," Casadevall chided, staring out at his audience at a midmorning talk on the subject, "you just had breakfast, and you're probably already thinking about lunch." (My stomach grumbled agreeably.) A crowd of crocodiles wouldn't have had to think about food for a week, he pointed out. And yet after the dinosaurs went extinct, their fellow reptiles did not stage a second act; diminutive, inefficient—but fungal-pathogen-free—mammals rose to dominance instead.

Warm-bloodedness would have provided especially critical protection against pathogens during *Homo sapiens'* early years on Earth. Most of our pathogens back then were adapted to the ambient temperature because they lived at least part of their lives in the environment. (There weren't enough of us around back then for them to make a full-time living in our bodies.) Keeping our blood warm foiled them. Unfortunately, today most of our pathogens come from other mammals, which means that by the time they get to us they've already adapted to warm blood. Still, we run fevers to scorch them out anyway, Casadevall points out, an atavistic gesture to an earlier era when our interior heat saved us from pathogens.

The problem is that our warm blood repels fungal pathogens only because its temperature diverges from the ambient temperature around us, to which fungi are accustomed. If fungal pathogens evolved to tolerate higher temperatures, that gradient would disappear. It is technically possible: in lab experiments, fungi that usually perish at temperatures above 82° Fahrenheit can be bred to tolerate temperatures up to 98°. Climate change could produce the same result, on a planetwide scale, slowly but inexorably training fungi to tolerate increasing temperatures, including, at some point, the heat of our blood.

Heat-tolerant fungi, if they emerge, would pose an infectious disease threat like no other, Casadevall says. Except for warm-bloodedness, we have no defenses against them. "If you don't believe me, ask the amphibians," he says, referring to the fungal pathogens that have decimated them. "Ask the bats."[54]

As the global temperature rises, fungal pathogens have already started to creep into the infectious-disease landscape. In California and Arizona, the soil-dwelling fungus *Coccidioides immitis* and *C. posadasii* caused seven times more human infections—dubbed "Valley Fever"—in 2009 than in 1997.[55] Disease-monitoring programs such as Health-Map and Pro-MED increasingly carry reports of outbreaks of fungal diseases. HealthMap reported twice as many in 2011 than in 2007, and Pro-MED reported seven times as many in 2010 than in 1995.[56] These could be random peaks that will be followed by dips—or they could be harbingers of a wave of climate-change-driven fungal pathogens to come.

⊙

Climate change, like all of the other ways we've put ourselves at risk of pandemics today, is a product of modernity. We can trace each excess atom of carbon in the air today to specific activities that occurred with the rise of capitalism, from the firing up of the first coal-powered factories to the gas-guzzling cars and jets of today. That suggests that tackling the next pandemic requires grappling, in one way or another, with the novel problems created by industrialization and globalization. But that would solve only part of the problem. Tomorrow's pandemic may be a product of modernity, but pandemics in general are not. In fact, the specter of contagion has been haunting our kind for millions of years.

While the dynamics of infectious disease, from that of cholera in the nineteenth century to the emerging pathogens of today, are dictated by specific historical conditions, our modern confrontation with pathogens is just the latest skirmish in a much longer, more fraught, more complex confrontation between us and microbes.

NINE

THE LOGIC OF PANDEMICS

There's no straightforward record of the ancient pandemics that plagued us. They can be discerned only obliquely, by the contours of the long shadows they've cast. But according to evolutionary theory and a growing body of evidence from genetics and other fields, pandemics and the pathogens that cause them have shaped fundamental aspects of what it means to be human, from the way we reproduce to the way we die. They shaped the diversity of our cultures, the outcomes of our wars, and lasting ideas about beauty, not to mention our bodies themselves and their vulnerability to the pathogens of today. Their powerful and ancient influence informs the specific ways modern life provokes pandemics the way the tides shape the currents.

Disease is intrinsic to the fundamental relationship between microbes and their hosts. All it takes to confirm that is a brief tour through the history of microbial life and a peek inside our own bodies. Humans dominate the planet in modern times, but in the past, it was the microbes that ruled. By the time our earliest ancestors, the first multicellular organisms, clambered out of the sea around 700 million years ago, microbes had been colonizing the planet for nearly 3 billion years. They had radiated into every available habitat. They lived in the sea, in the soil, and deep inside Earth's crust. They could withstand a wide range of conditions, from temperatures as low as 14°F to as high as 230°F, feeding on everything from sunlight to methane.

Their hardiness allowed them to live in the most extreme and remote places. Microbes colonized the pores inside rocks, ice cores, volcanoes, and the ocean's depths. They thrived in even the coldest and saltiest seas.[1]

For the microbes, our bodies were simply another niche to fill, and as soon as we formed, they radiated into the new habitats our bodies provided. Microbes colonized our skin and the lining of our guts. They incorporated their genes into ours. Our bodies were soon home to 100 trillion microbial cells, more than ten times the number of human cells; one-third of our genomes were spiked with genes that originated in bacteria.[2]

Did our ancestors willingly play host to the intruding microbes that colonized their interiors? Possibly. But probably not. For like the outsized military of an insecure state, we developed a swollen arsenal of weaponry to surveil, police, and destroy microbes. We shed layers of skin to rid ourselves of the microbes that would colonize its surface. We constantly blinked our eyelids to wash microbes off our eyeballs. We produced a bacteria-killing brew of hydrochloric acid and mucus in our stomachs to repel microbes that might attempt to colonize their interiors. Every cell in our body developed sophisticated methods of protecting itself from microbial invasion, and the capacity to kill itself if it failed. Specialized cells—white blood cells—coursed through our bodies with no other role but to detect, attack, and destroy intruding microbes. In the time it just took you to read these lines, a flood of them washed through your entire body, surveilling for signs of microbial invasion.

The development of these immune defenses attests to the ongoing threat that microbes must have posed. To survive, our bodies had to be finely tuned to fight contagion. Our immune defenses were not some vestigial backup system, like a retired security guard relaxing in the back of a rarely visited shop. They were ever alert and activated by hair trigger. Today, just looking at a picture of a person suffering a microbial attack, such as someone with skin lesions or who is sneezing, will cause our white blood cells to start pumping out elevated quantities of immune fighters such as cytokine interleukin-6, just as if we'd been invaded by a microbe ourselves.[3]

Sustaining this battle readiness against microbes wouldn't have been easy. Anytime our immune system was activated, we needed to consume more oxygen. During periods when we had to expend energy elsewhere, such as when our bodies were incubating and nursing our young, we were forced to let down our guard. Then as now, we lacked sufficient resources to keep the expensive immune system running. Protecting the body from the appetites of microbes is, in the lingo of biologists, "costly." We paid the price because surviving in a microbial world required it.[4]

But even though the immune system helped us ward off pathogenic incursions into the body, it did not seal us off entirely. Far from it: to this day, any diminishment in our battle readiness—or change in microbes' ability to foil our defenses—results in violent confrontation. When our immune defenses are weakened by age or disease or exhaustion, microbes invade our cells. Once they do that, they wreak havoc in a variety of ways. Some replicate with abandon, using up our nutrients or damaging our tissues in the process. Others, like cholera, secrete toxins that help them replicate or spread. Some simply provoke reactions from other sensitive bodily systems. Their methods vary, but the result is the same: they thrive while we sicken.

We call these culprits "pathogens," but really they are just microbes doing the same thing they do everywhere else: feeding, growing, and spreading. And they do so relentlessly. That's their nature. Under optimal conditions, microbes double their numbers every half hour. They never age. So long as there's enough food around, they won't die unless something kills them. That is to say, they will predictably exploit the resources available to them to the fullest extent possible. If that results in epidemics and pandemics, so be it.

The logic of microbial life and the nature of our immune defenses conjure up a pandemic-scarred past. But there's more. Evolutionary biologists and geneticists interpret certain anomalies, such as unusual signatures in our DNA and strange behaviors that are otherwise difficult to explain, as clues, too. For a growing number of experts, they're as suggestive as the trembling hands of a seemingly unscathed trauma victim would be to a criminal detective: only a violent, pandemic-plagued past could explain them.

⊙

These anomalies are not what most people would consider either strange or hard to justify. They are two fundamental parts of our life cycle: sexual reproduction and death. We take them as given. But for evolutionary biologists, they are puzzling developments in our evolution that demand explanation.

Grasping this rather counterintuitive notion requires a brief digression into what's called the "selfish gene theory" of evolution. The basic idea is that genes—or, rather, the genome, which is the entire complement of genes in a given individual—are the movers and shakers of evolution. The genome consists of long twisted molecules of DNA (or RNA), which are carried around in each of our cells, bits of which (the genes) provide instructions for a wide range of biological traits, from eye color to nose shape to the sound of one's voice. According to the selfish gene theory, all of evolution can be boiled down to their machinations. Some genes, by dictating or "coding" for traits that help them spread more widely, become dominant. Others, which code for traits that are useless or harmful to their own dissemination, die out.

It is in light of selfish gene theory that sexual reproduction and death come to seem confounding, for neither sex nor death is a particularly efficient means of disseminating genes, given the alternatives.

Consider sexual reproduction. At one time, all life on the planet reproduced asexually (by cloning or other methods). There were no sexually reproducing creatures. But at some point in the history of evolution, sexual reproduction emerged. And yet, from the point of view of our genes, it was a vastly inferior strategy compared to other methods of reproduction. Organisms that clone themselves pass on 100 percent of their genes to their offspring. Sexual reproducers must not only partner up with another individual to reproduce, but both parents lose half of their genes in the bargain, for the resulting child inherits half of its genes from each parent.

To survive, the first sexually reproducing organisms would have had to outcompete the cloners, who dominated the resources and habitats of the planet. But how could they? In the 1970s, the evolutionary biologist William Hamilton created a computer model to simulate what that contest looked like. The simulation set up a population in which half of the individuals practiced cloning and half paired up and had sex. (Imagine a clan of all-female Amazons who replicate themselves without males, alongside a tribe of females who could reproduce only with the help of a male.) Everyone was equally subject to the kind of random deaths that befall populations in the wild, like being attacked by predators or frozen in an ice storm. The model then calculated the reproductive success of the two tribes, counting the number of offspring they each produced.

The cumulative effect of the two different reproductive strategies didn't take long to reach its logical conclusion. Every time Hamilton ran the model, the sexual reproducers rapidly went extinct. Random deaths in the sexually reproducing tribe resulted in a disproportionate loss to the mating pool (as anyone who has tried to find a date after the age of forty can understand intuitively). Not so for the cloners, who maintained their vigorous rate of replication regardless of random losses. It didn't matter that the offspring of the sexually reproducing tribe were more genetically diverse and therefore more resilient to long-term changes in the environment. The burden of random deaths was too immediate to allow those benefits to manifest themselves.

Thus sexual reproduction should have been an experiment that failed. And yet it didn't. Eventually, the reproductive strategy of our most distant ancestors spread throughout the animal kingdom, including in us, many years later, in whom it became a central preoccupation.

It was Hamilton who offered a startling explanation that solved the mystery: sex evolved, he said, because of pathogens.

Sexual reproduction requires a profound genetic sacrifice, he noted, but the payback is that the offspring of sexual reproducers are genetically distinct from their parents. That was no big advantage in surviving hostile weather or predators, Hamilton observed, but it was a huge advantage in surviving pathogens. That's because pathogens, unlike the weather or predators, refine their attacks upon us.

Imagine a pathogen that first strikes when you're a baby. As you develop, the pathogen goes through hundreds of thousands of generations. By the time you're an adult (if you've survived the ravages of the pathogen) and are ready to reproduce, the pathogen has become far better at attacking you than you are at defending yourself against it. While your genetic makeup has stayed the same, the pathogen's has evolved.

But individuals who clone themselves provide pathogens exact replicas of the target they've already gotten so good at stalking. They endow their offspring with the worst possible chances of surviving the pathogen's appetites. Much better, in that case, Hamilton theorized, to produce offspring that are genetically distinct from you, even if that means for-saking half of your own genes.

Scientists have shown how refined pathogens' attacks become over time by experimentally transferring the pathogens of an old individual into a young one. One study cited by the evolutionary zoologist Matthew Ridley focused on long-lived Douglas fir trees, which are routinely at-tacked by scale insects. (Although scale insects are not microbes, they are disease-causing organisms just like microbial pathogens.) In the wild, old trees are more heavily infected than young ones. This is not because the old trees are weaker than the young ones, as one might think. The old are more heavily infested because their pathogens have had more time to adapt to them. When scientists transplanted the scale in-sects of an old tree onto a young tree, the young tree suffered the same heavy burden of disease as its elders. It's easy to see how, with pathogens like that around, sexual reproduction would provide a better chance at survival than cloning.[5]

Since Hamilton first articulated his theory about pathogens and the evolution of sex, a large body of supportive evidence has accumulated. Biologists have found that species that practice both sexual and nonsex-ual reproduction will switch between the two depending on the presence of pathogens. When raised in a lab devoid of their usual pathogens, or in the presence of pathogens that are altered in such a way that they can-not evolve, the roundworm *Caenorhabditis elegans* will mostly replicate without sex. But when stalked by pathogens, it will reproduce sexually instead. In other experiments, scientists altered roundworms so that they cannot sexually reproduce. When they reared these worms with patho-gens, the nematodes went extinct within twenty generations. In contrast, when they allowed roundworms to practice sexual reproduction, they survived alongside their pathogens indefinitely. Withstanding pathogens seems to require the special benefits provided by sexual reproduction.[6]

By forcing the evolution of sex, pathogens may have forced an ad-ditional adaptation: death. The notion that death is some optional thing that can "evolve" may seem counterintuitive. The idea that deteriora-tion and death are inevitable is central to how most of us think about life. We think of the body as a kind of machine that inevitably wears out over time. Individual parts fail and the damage accumulates. Finally, after some critical threshold is passed, the entire machine stops work-ing. Thus we say that nobody can "cheat death." We even equate the word "aging," which is simply the passing of time, with diminishment. (What we really mean is what biologists call "senescence," a gradual deterioration of function that proceeds with the passing of time and ultimately leads to death.)

But senescence and death are not inevitable facets of life. There are examples of immortality all around us. Microbes live forever. Trees don't deteriorate with time. On the contrary, as they age they get stronger and more fertile. For microbes and many plants, immortality is the rule, not the exception. There are even some animals that don't age: clams and lobsters, for example. Death, for them, is caused solely by external factors, not internal ones.

One way the human body is distinctly different from a machine is that it can repair itself. After we exercise, we repair ourselves from the damage we've inflicted to our muscles. When our bones are broken or skin ruptured, we grow new bone tissue and new skin. (There are even reports of people regrowing severed fingers.)[7] Our cells have a wide range of ways they repair themselves from insult. Other animals have this capacity to self-repair. Worms rebuild their severed wriggling bodies. Starfish regrow their arms. Lizards regrow their tails. Such repairs actually make us stronger, not weaker.

Scientists have found that far from being an inherently inevitable process, senescence is controlled by particular genes, variously called "suicide genes" or "death genes." Their job is to progressively turn off the processes of self-repair that keep our bodies in good condition. They're like a host switching off the lights at the end of a party. It happens at a certain time, no matter what.[8]

The discovery of these genes dates back to the 1970s, when scientists found that removing certain glands from a female octopus could postpone her otherwise inevitable death. Normally, a female octopus will stop eating and die, like clockwork, ten days after she finishes tending her eggs. But surgically removing the glands that control maturation and breeding resulted in an octopus that behaved quite differently. After laying her eggs, she resumed eating and survived for another six months.[9] Scientists have similarly pinpointed genes with no known purpose other than to trigger deterioration and death in worms and flies. When those genes are experimentally inactivated, death is delayed. The worms and flies live on.[10]

So far, it seems unlikely that genes with such singular purpose will be found in people. More likely, suicide genes in humans play a number of different roles, both beneficial and detrimental. Genes that control inflammation may protect us from wounds and infections when we're young, but then go suicidal and start attacking healthy cells. The conditions that trigger these abrupt about-faces have yet to be pinpointed, but for obvious reasons they are the focus of much zealously followed research conducted by antiaging scientists.[11]

The discovery of suicide genes begs the same question that sex does. How would such genes have ever evolved? The programmed death such genes cause is a loser compared to the alternative. In a straight evolutionary contest, individuals encumbered by suicide genes, collapsing halfway to the finish line while their rivals surge ahead, would surely lose. There would have had to have been some immediate payback to compensate for such a heavy debility.

That payback, according to what's called the "adaptive theory of aging," is protection against species-leveling pandemics. Immortality undoubtedly has its perks, but there are also significant drawbacks. One is that immortal species tend to rapidly expand their numbers to the limits of whatever resources are available in the environment. That leaves them vulnerable to catastrophic events, like famines and pandemics, which can strike them in one fell swoop, killing everyone all at once.

We know that catastrophic events like this happened in the past with some frequency. After all, 99.9 percent of all the species that have evolved on Earth are now extinct. Those of us who remain today are the few survivors on our volatile planet. How did we do it? Immortal species, like microbes, were probably resilient against catastrophic famine and pandemics because they also practiced cloning. That meant that even a pandemic that wiped out 99.9 percent of their population wouldn't force them into extinction, for a small number of survivors could rebuild their numbers. But the odds for any immortal, sexually reproducing species would have been grim. One group of conservation biologists estimated the minimum population size required for most sexually reproducing animal species to remain viable to be around five thousand.[12] Others put the range between five hundred and fifty thousand, depending on the species. Any pandemic (or famine) that wiped out more than that number would extinguish a sexually reproducing species forever.[13]

The adaptive theory of aging posits that this was the context in which suicide genes evolved. The scenario would have gone something like this. Imagine two competing groups of sexually reproducing organisms. In one group, all are immortal. In the other, suicide genes have emerged and so some individuals slowly age and die. The first group is like a dense forest; the second is like a regularly culled one. When the pandemic arrives, the former group will fare as poorly as the dense forest does in a forest fire. The latter group will be more likely to survive, allowing suicide genes to spread.

Suicide genes obviously don't protect us entirely from the risk of famine and pandemic. But because individuals in our groups regularly age and die, "a little bit at a time," as the antiaging researcher Joshua Mitteldorf puts it, the risk that those events will cause an extinction is much lower. We age and die, Mitteldorf contends, as a sacrifice to pandemics.[14]

Both Hamilton's theory about the evolution of sex and the adaptive theory of aging are versions of what's called the "Red Queen Hypothesis," which has revolutionized modern biology. It's named after a scene in Lewis Carroll's *Through the Looking Glass*. Alice collapses to the ground after a vigorous bout of running with the Red Queen, only to find they'd made no progress at all. "You'd generally get to somewhere else," Alice says, "if you ran very fast for a long time, as we've been doing." The Red Queen explains the logic of why they did not: "Now, here, you see, it takes all the running you can do, to keep in the same place. If you want to get somewhere else, you must run at least twice as fast as that!"

What does this mean for our epidemic past and future? According to classic natural selection theory, as articulated in 1859 by Charles Darwin and as taught in high school biology classes around the world, pathogens and their victims adapt to each other over time, evolving toward a less fractious relationship. The Red Queen Hypothesis says otherwise. For every adaptation on the part of one species, it holds, there's a counteradaptation on the part of its rival. What that means is that pathogens and their victims don't evolve toward greater harmony: they evolve increasingly sophisticated attacks on each other. They're like spouses in a bad marriage. They run "very fast and for a long time," but they don't "get to somewhere else."

And that leads to the same conclusion as arguments about the nature of microbes and the immune system and the evolution of sex and death. That is, that the relationship between pathogens and their victims does not evolve toward greater accommodation. On the contrary, it's a continuous battle in which each side evolves increasingly more sophisticated ways to crack the other's defenses.

This suggests that epidemics are not necessarily contingent on specific historic conditions at all. Even in the absence of canals and planes and slums and factory farms, pathogens and their hosts are locked in an endless cycle of epidemics. Far from being historical anomalies, epidemics are a natural feature of life in a microbial world.

⊙

These theories about sex, death, and pathogens were not formulated to reveal the extent of our long entanglement with pathogens. They were attempts to resolve theoretical problems in natural selection, the cornerstone theory of modern biology. But strange patterns in our genes, and the way geneticists and other scientists have tried to understand what they mean, support their theoretical claims.

One of those strange patterns has to do with the nature of genetic diversity among us. Colloquially, we say that each of us is "genetically unique," but that's actually not accurate. In fact, we all have the same genes. Each of us has genes that tell the body how to build a nose or how to shape an ear, for example. (A gene is simply a specific segment of DNA where instructions for specific traits are stored.) What we have among us are different variants of the same gene, for the sequence of chemicals in that segment varies from individual to individual. Your variant, for example, may call for an attached earlobe and mine for a hanging one.

Sex and mutations introduce new variants and combinations into our genomes at a regular clip. But it's a messy process with no direction. It's like blindly throwing a wrench at a bicycle. Most of the time, new variants are downright unhelpful. The genome is degraded by the variant just as the bicycle would be. Sometimes, the variant is neutral and there's no noticeable effect at all. Very rarely, a random variant will happen to coincide with events that make it useful. Over time, unhelpful genetic variants are methodically weeded out while the beneficial ones come to dominate. And so, when geneticists compare the genomes of a bunch of different people at a given moment in time, they expect to find a certain degree of genetic variation, but not a huge amount.

And yet, when geneticists zoom in to one part of the genome, they find a singular anomaly. It's the part of the genome where certain pathogen-recognition genes lie. These genes provide instructions for building what are called human leukocyte antigens (HLA), proteins that signal to the immune system when a cell has been infected. (They do this by binding to a fragment of the pathogen and displaying it on the surface of the cell, like a flag.) On this one part of the genome, we've maintained a huge number of variants among us. Our HLA or pathogen-recognition genes are more diverse, in fact, than any other part of the genome, by two orders of magnitude. So far, more than twelve thousand variants have been discovered.

There are two possible explanations for it. Either each of those twelve thousand variants is neutral and thus the variation is meaningless—which is hard to believe given the sheer number of variants—or some powerful force has reversed the normal pressures that reduce variation, making it somehow advantageous for us to maintain a vast library of old genetic variants.

That force may be pathogens causing repeated cycles of epidemics. To cause repeated epidemics in the same population, a pathogen must switch between different strains to evade detection, like a thief using different disguises to repeatedly rob the same bank. Retaining a large number of pathogen-detection genes among us ensures that there'll always be a few individuals who can suss out the latest disguise. Each pathogen-detection gene variant thus neither fully dies out nor sweeps into dominance. We carry them around with us, like a treasure chest full of specialized detection tools handed down from generation to generation.[15]

What's more, we've been doing this for millennia. We have a lot of old genes in our genomes, genes for useful traits like eyes and brains and backbones, which we share with other species. Our pathogen-recognition genes are on par with these. Some of the pathogen-recognition genes embedded in modern people's genomes are 30 million years old. They've survived among us even as we've split off into different species multiple times. That suggests that pathogens have been cyclically causing epidemics, dying down, then lashing out again for geological eons.[16]

⊙

Our genomes also contain clues about a specific pandemic in our past. This one struck the hominid line (of which *Homo sapiens* are the sole survivors) around 2 million years ago. The evidence lies in a gene that controls the production of a particular compound called a sialic acid. Over the course of three hundred thousand years—a heartbeat in evolutionary time—every individual who produced this sialic acid died out or failed to reproduce, leaving behind only those who didn't produce the sialic acid, because they had variant of the gene that inactivated it.

What could have caused such a dramatic change so quickly? The scientist who discovered the loss of the gene, the sialic-acid expert Ajit Varki, suspects that it was a pandemic. That's because, besides a variety of other roles in cell-to-cell interactions, sialic acids are used by pathogens to invade cells. (They bind to them, which is like turning a key in a lock, allowing them access to the interior of the cell.) A pandemic caused by a pathogen that invaded cells using the particular sialic acid that was lost could have killed off all the individuals who produced it, leaving only those who didn't. Varki suggests that it was probably some form of malaria, noting that the malaria parasite *Plasmodium reichenowi*, which today causes malaria in chimpanzees, binds to the lost sialic acid, which is called N-Glycolylneuraminic acid, or Neu5Gc.[17]

That malaria-like pandemic had profound consequences for the survivors. Their cells, unlike those of every other primate and all other vertebrates, no longer produced Neu5Gc. That meant that any attempt at conception between a survivor and anyone who hadn't lived through the pandemic would have failed. The survivor's immune system would register Neu5Gc-laden sperm cells, or those of a developing fetus, as foreign and attack them; as Varki's experiments on genetically engineered mice have shown, survivors could reproduce only with each other.

A new species would have been born. Indeed, according to fossil evidence, the first upright, walking hominid species, *Homo erectus*, diverged from their predecessors, the ape-like *Australopithecus*, right around the time when New5Gc was lost. If Varki is right, our first pandemic helped make us human.[18]

The striking thing about these findings about ancient pandemics is that the paradoxical observations they're based on were made in the course of unrelated inquiries. Both the discovery of our lost sialic acid and that of the diversity in our pathogen-recognition genes were flukes. Varki discovered the lost sialic acid in 1984, when he administered horse serum to a patient with bone-marrow failure and found that the patient's immune system reacted to the sialic acids in it. He spent decades figuring out why, stumbling upon the story of the ancient pandemic in the process. Scientists discovered the diversity in our pathogen-recognition genes in the course of attempting organ transplants. Unless the donor and recipient shared identical pathogen-recognizing HLA genes, surgeons found, the recipient's immune system would attack the donor's organ as if it were pathogenic. Attempts to match donors and recipients according to their HLA genes slowly revealed the vast scale of variation among us. And yet despite the happenstance nature of these discoveries, both led to conclusions that jibed with the theories of evolutionary biologists, separately attempting to resolve their own paradoxes. We'd probably know even more about our pandemic past if we actually attempted to search for it on purpose.[19]

⊙

While the tracks that ancient pandemics left are faint, at least for now, their aftershocks are not. They can be felt by all of us, from the idiosyncrasies in our immune systems to the historical trajectories of our ancestors, in ways that scientists are just now starting to understand.

Ancient epidemics led to the development of our heightened immune responses. These now predispose us to a range of ills, including spontaneous abortions. Five percent of all women experience recurrent, spontaneous abortions due to immunological reasons: in one way or another, the mother's immune system, spuriously sensing a foreign intruder, attacks the fetus. Our bodies respond similarly to any tissues and cells of fellow *Homo sapiens*. That's why, unless the immune systems of transplant recipients are medically suppressed, they will almost certainly attack donor organs (besides those donated by an identical twin).[20]

Our heightened immune responses, in particular those we developed to survive the ancient pandemic that Varki discovered, may predispose us to developing cancer, diabetes, and heart disease if we consume red meat. Red meat, being the flesh of mammals, is rich in Neu5Gc, the sialic acid we lost. Consuming it may trigger the same kind of immune reaction in us that mating with *Australopithecus* did among our ancestors 2 million years ago. Our bodies, registering their tissues as foreign and pathogenic, attempt to fight them off with inflammation. Those tiny inflammatory responses, over time, may increase the risk of developing cancer, heart disease, and diabetes, all of which have been linked to inflammation. In lab experiments, Varki found, mice genetically engineered to react to Neu5Gc with inflammation as we do suffer a fivefold increase in cancers when exposed to the sialic acid.[21]

Genetic variants that helped us survive pathogens in the past now burden us with heightened risks of contracting other diseases and conditions. The most famous is the sickle-cell gene, which deforms red blood cells. This gene spread among people in sub-Saharan Africa who suffered malaria epidemics because it slashed the death rate from that disease. In 2010, more than 5 million infants were born with the gene. But while it helps them survive malaria, those born with a double dose of the gene suffer sickle-cell anemia, a disorder that is fatal in the absence of modern medicine.[22]

Similarly, the gene that helped Africans survive sleeping sickness now puts them at risk of kidney disease, which may explain the high rates of kidney disease in African Americans today.[23] Genetic changes that helped people survive malaria made them more susceptible to other pathogens such as cholera.[24] The genetic mutation that allowed people to survive leprosy, which is present in 70 percent of modern Europeans, is now associated with inflammatory bowel diseases such as Crohn's disease and ulcerative colitis. Other genetic mutations, which bestowed on Europeans greater protection against bacterial infections, simultaneously damaged their ability to digest gluten. The result is celiac disease, which afflicts up to 2 percent of European populations today.[25]

Genes that studded our red blood cells with proteins resulting in what is now known as the A and B blood groups, which may have evolved to help protect people from severe infections during pregnancy, now make people more susceptible to arterial and venous thromboembolism.[26] Particular variants in our pathogen-recognition genes, which protected us from ancient epidemics, correlate with a range of autoimmune disorders, from diabetes and multiple sclerosis to lupus.[27] Whether or not people survive HIV or malaria, or launch an adequate immune response to measles, depends upon their particular pathogen-recognizing HLA genes, which evolved to help ward off the pathogens of the past.

Ancient epidemics and pandemics have cast a long shadow upon us. While connections between our genetic adaptations to ancient epidemics and our vulnerability to modern pathogens have only recently been detectable, thanks to advances in genetic research, scientists expect that many more such connections exist and are yet to be found. It may be that much of our vulnerability to the pathogens of today—and tomorrow—is shaped by how our ancestors survived the pathogens of the past.[28]

⊙

Given the outsize role pathogens and pandemics have played in our evolution, it stands to reason that they've probably helped shape our behavior, too. According to psychologists, historians, and anthropologists, they have. The evolutionary psychologists Corey L. Fincher and Randy Thornhill theorize that culture itself—the differentiation of populations into behaviorally and geographically distinct groups—originated as a behavioral adaptation to an epidemic-filled past.

The theory starts with the idea of "immune behaviors." These are social and individual practices that help people elude pathogens, such as avoiding certain landscape features like wetlands or swamps, or practicing certain culinary rituals, like adding spices with antibacterial properties to foods. These behaviors are not necessarily purposely designed to protect people from pathogens; people may not have even been aware that they helped do so. But immune behaviors, once developed, stick around because the people who indulge in them are less vulnerable to infectious diseases. The behaviors, passed down through the generations, become entrenched.

In our early evolution, when human mobility was relatively limited, immune behaviors would have been highly localized, since pathogens and their victims would have been so intimately adapted to each other. Traces of this are detectable today. In Sudan, anthropologists have found that immune behaviors that protect people from a pathogen called *Leishmania* differ from village to village. This variability likely corresponds with the variability of the pathogens to which they're exposed: what works in one place against one strain of the pathogen may not work as well elsewhere against different strains. Indeed, more than one hundred genetically distinct strains of *Leishmania* have been found over small geographic distances.[29]

The specificity of immune behaviors would have made interactions with outsiders especially risky. Because they would not be privy to the specialized knowledge of local pathogens and the appropriate immune behaviors required to avoid them, they could undermine and disrupt these practices (or introduce nonlocal strains of the pathogens). Thus the value of insiders would have grown in relation to outsiders. With it, so did practices that highlight those differences—like costumes and tattoos—and attitudes that police them, like xenophobia and ethnocentrism. The result, over time, is the development of distinct cultural groups.

The disease historian William McNeill hypothesizes that these highly localized immune behaviors contributed to the development of the caste system in India, in which strict rules limit contact between castes, and there are elaborate rules for purifying the body if contact occurs. These may be the result, in part, of each group having specific immune behaviors tailored to its local pathogens, McNeill speculates, and the resulting need for a system that policed group boundaries.[30]

Suggestively, in places where there are more pathogens, there are more ethnic groups (among traditional peoples), and vice versa.[31] Of all the various factors that could potentially predict the level of ethnic diversity in a given region, pathogen diversity is one of the strongest.[32] And in experiments, people who are made more aware of pathogens express greater allegiance to their ethnic group, suggesting that biases toward one's own group, the basis of cultural difference, is indeed linked to fear of disease. In a 2006 study, anthropologists found that eliciting people's fears of contagion (for example, by indicating that a glass of milk they were about to drink was spoiled) heightened their ethnocentrist attitudes, compared to people whose fears of contagion were not thus heightened.[33]

The differentiation of cultural groups by pathogens also dictated the outcome of confrontations between them. Groups of people have been able to vanquish other groups by wielding what McNeill calls an "immunological advantage." They simply introduce pathogens to which they've adapted but against which their rivals have no immunity. It happened in West Africa three thousand years ago, when Bantu-speaking farmers who'd adapted to a deadly form of malaria penetrated the interior of the continent, bringing the pathogen with them. They rapidly defeated the hundreds of other linguistic groups believed to have populated the region in what historians call the "Bantu expansion." Immunological advantages allowed the people of ancient Rome to repel invading armies from northern Europe, who perished from the Roman fevers to which locals had adapted. The protection afforded by Rome's immunological advantage rivaled those of a standing army. "When unable to defend herself by the sword," the poet Godfrey of Viterbo noted in 1167, "Rome could defend herself by means of the fever."[34]

Most famously, Europeans conquered the Americas starting in the fifteenth century by decimating native peoples with the Old World pathogens to which they had no immunity. Smallpox introduced by Spanish explorers killed the Incas in Peru and nearly half the Aztecs in Mexico. The disease spread throughout the New World, destroying native populations ahead of European settlement.[35] Meanwhile, the people of tropical Africa repeatedly repelled the forays of European colonizers, who were felled by the malaria and yellow fever to which locals had adapted. (One unhappy result was the development of the brutal Atlantic triangle trade of the sixteenth to nineteenth centuries. Having failed to establish colonies in sub-Saharan Africa, Europeans carried captives from Africa across the ocean to the Americas to serve as slave labor on their sugar plantations.) These and other confrontations, decided by the immunological distinctions among us, continue to reverberate through modern society today.[36]

⊙

Seemingly contradictory ideas about beauty, in particular the attractiveness of potential mates, may have evolved as immune behaviors, too. While the precise architecture of romance remains decidedly mysterious, evolutionary biology suggests a few general rules. One is that people should be attracted to mates who will be good coparents and help them produce viable children. That's just simple logic: people attracted to bad coparents tend not to have many kids, or not many who survive, diluting their numbers over time.

The contradiction is that in the case of humans, the attractiveness of mates doesn't seem to correlate with their likelihood of being good coparents. Cross-cultural studies have shown that women find male facial features that are controlled and made more pronounced by the hormone testosterone—broad chins, deep-set eyes, and thin lips—attractive. In general, the more testosterone a male has, the more likely he is to be attractive to women.[37] And yet, at the same time, the more testosterone a male has, the less likely it is that he will be a good coparent. Compared to low-testosterone males, high-testosterone males are more likely to engage in antisocial behavior and are less likely to get married. If they do marry, they are more likely to get divorced, have extramarital affairs, and act violently toward their spouses. A high level of testosterone, in that case, should make males *less* attractive to females. But it's just the opposite.[38]

Broad chins and deep-set eyes, in other words, are like the peacock's tail. Long, heavy, and conspicuously showy, the peacock's tail is a clear hindrance to male birds' survival. Female peahens looking for good mates should prefer male birds with less showy tails. But numerous studies have shown that, like human females who prefer high-testosterone males, peahens prefer male birds with the longest, fanciest tails.

The reason why, evolutionary biologists say, is that a peacock's long, fancy tail—precisely because it is a hindrance—signals to the peahen that he is a strong, able mate. It's advertising. And one thing it advertises is the strength of the peacock's defenses against pathogens. Peacocks with the longest, fanciest tails, scientists have found, have stronger immune systems and are less pathogen-infested than peacocks with shorter tails. And choosing them over peacocks with short, dull tails does help peahens enjoy greater reproductive success. Peahens who mate with long-tailed peacocks have bigger babies at birth who are more likely to survive in the wild, compared to peahens who mate with shorter-tailed peacocks. And so despite the fact that their dazzling, elaborate tails are a hindrance to peacocks' survival, peahens continue to find them attractive.

The male features in humans that indicate high-testosterone levels may perform a similar function. They, too, advertise the strength of a male's immune system: high levels of the hormone correlate with stronger immune defenses. It may be that females find high-testosterone facial features attractive for the same reasons that peahens find long, showy tails attractive: they demonstrate the pathogen-fighting prowess of their mates.

In a study of twenty-nine different cultures, psychologists found that those that placed more emphasis on the physical attractiveness of potential mates were indeed those with higher burdens of pathogens. Another study found that females who express greater awareness of contagion prefer males with more masculine features. There's also experimental evidence to support the link between ideas about male beauty and contagion. Scientists experimentally manipulated subjects' fear of contagion (for example, by showing them pictures of white fabric stained with blood) and then asked them to judge male features. They found that women whose awareness of pathogens had been heightened preferred images of males with more masculine features, compared to women who were not thus provoked.[39]

Another curious facet of attractiveness and mate choice that may have originated as a strategy to survive ancient epidemics has to do with pathogen-recognizing HLA genes. Choosing a mate with pathogen-recognition genes different from your own improves the chances that your children will be able to survive a broad range of pathogens. Indeed, couples whose pathogen-recognition genes differ enjoy greater reproductive success than couples whose pathogen-recognition genes are more similar. (They suffer fewer spontaneous abortions and their children are more closely spaced in age, suggesting that they experience few miscarriages.)

Of course, the composition of other people's pathogen-recognition genes can influence our choices about mates only if we can somehow distinguish between people with similar pathogen-recognition genes and those with exotic pathogen-recognition genes. Although most people are unaware of it, it turns out that we can. Numerous studies have shown that people, like other animals, can sense the composition of others' pathogen-recognition genes by scent. (Precisely how pathogen-recognition genes influence body odor is unclear. It may revolve around how the proteins coded by the genes bind to cells or affect the bacterial fauna in the body that create odors.) And people have preferences based on those odors. In one study, subjects whose pathogen-recognition genes had been typed were asked to wear cotton T-shirts for two nights in a row (while refraining from using perfumes in soaps or other products and eating foods that produced strong odors). The T-shirts were then stuffed into unlabeled jars, which were presented to the subjects to sniff.

Each preferred the scent of those T-shirts worn by people whose pathogen-recognition genes differed the most from their own.[40]

That's not to say we choose mates based solely, or even in part, on their body odor, of course. But it's quite possible that we had to in our epidemic-plagued past. To this day we can sniff out the difference and feel a twinge of residual desire based on it.

Microbes have exerted a similarly powerful influence on us via their perch from inside our bodies. Scientists are just starting to unravel the mysteries of the microbes that live in and on us, collectively known as the microbiome. So far, they've found that they're often invisible puppet masters, too, with critical processes such as that of brain development in mammals, sex in insects, and immunity in mice triggered solely by the presence of certain microbes.[41] The microbes that live in human guts influence our risk of developing obesity, depression, and anxiety. They may play a role in controlling our behavior as well. Experimentally ridding mice of their microbes altered their behavior in suggestive ways, reducing both their anxiety responses and ability to perform tasks requiring memory; exposing one mouse to the microbes of another led it to behave in ways that mimic the other.[42]

All of which is to say that our vaunted sense of individuality is an illusion. Animals like us, as the evolutionary biologist Nicole King has said, have never been single organisms. For better or worse, we're "host-microbe ecosystems." Microbes shape us from without and also from within.[43]

That is to say, pathogens and pandemics are not solely the products of modern life. They're part of our biological heritage. The predicament we find ourselves in today, on the threshold of a new pandemic, is hardly exceptional. It's of a piece with hundreds of millions of years of evolution.

⊙

In many ways, we remain as diminished by pathogens today as we were eons ago. Globally, we've conquered barely more than a handful. New pathogens encroach upon us by the hundreds, threatening a pandemic. Meanwhile, old ones continue to exact their pounds of flesh: nearly half of all deaths in people under the age of forty-five are due to infectious disease.[44]

And yet at the same time, our prospects have never been better.

Consider the fact that of the three existential challenges faced by all species, pathogens are just one. Our conquest over the other two—predators and Earth's often hostile climate—has been nearly complete. We've been incrementally transforming hostile climes to suit our needs and comforts since our ancestors tamed fire one million years ago, banishing the night and the cold the way our central heating systems and hermetically sealed glass windows do today.[45] Our battle with predators was settled when we walked out of Africa one hundred thousand years ago and into the world's continents, rapidly driving every other large mammal—and the predators that hunted them—into extinction. We rid our habitats of the American lions, the mastodon, the mammoths, the saber-toothed cats, and the other hominids such as Neanderthals that might have preyed upon us. Our sole predators left now are other humans.[46]

I don't mean to argue that the way we subdued the environment and other species was without negative ramifications. But it shows the extent of human capacities when we can apply our intelligence and tool-making skills to the task. Because both of these existential challenges have been obvious to us for thousands of years—even our earliest ancestors could recognize the destructive power of storms and the danger that predators posed to them—we've been able to wield our own agency to overcome them.

In contrast, for most of our history, we've been unaware of pathogens' role in our lives. We developed the technology to detect microbes less than two centuries ago. We are just beginning to grasp the extent of their secret world today. It may have seemed, with the development of antibiotics and other wonder drugs in the mid-twentieth century, that we'd conquered our old foes. But seen in a larger historical context, we look more like climbers on top of a foothill, mistakenly thinking we've reached the summit. The project of applying our intelligence and tool-making skills to the challenge pathogens pose has only just begun.

TEN

TRACKING THE NEXT CONTAGION

Well, you've scared the shit out of me," the bearded man sitting in the front row said to me when the question-and-answer session began.

It was the spring of 2015 and I'd just finished an hour-long presentation about pandemics to a crowd of students and faculty at a small boarding school outside Minneapolis. It wasn't the first time I'd heard a version of that response. I'd been talking about the science, politics, and history of pandemics to physicians' groups, students, and academics for a year or so. And afterward, as audience members filed out of the auditoriums and conference rooms, I had overheard more than a few laughing nervously and whispering to each other about needing to go wash their hands.

Remember the hysteria around SARS? the bearded man asked. Avian flu? Ebola? We'd panicked every time, but then when the outbreaks subsided, we went back to ignoring contagions altogether, he said. What's the point? Would we ever be able to terrorize ourselves into containing the next contagion?

From his perspective, I had just spent the previous hour under stage lights frightening my audience, so my response probably seemed contradictory. I agreed with him. Outbursts of fear are indeed futile. But that's not because fear itself is a problematic response to the challenge that pathogens present to us, which is formidable. It's because of where that fear comes from.

⊙

One of the most spectacular demonstrations of fear about pathogens in recent history occurred during the 2014 Ebola epidemic in West Africa. While scores of impoverished slum dwellers in Monrovia and Freetown perished, people from the suburbs of Kentucky to the air-conditioned offices of Canberra were gripped by the terrifying notion that Ebola would come to get them, too.

According to polls, nearly two-thirds of Americans feared an Ebola epidemic in the United States.[1] Terror-struck, they avoided contact with anyone who'd visited any locale on the same landmass as the affected countries, no matter how distant. From Connecticut to Mississippi, schools closed their doors to teachers and students who'd visited countries that were thousands of miles away from the outbreak—Kenya, South Africa, Zambia, Rwanda, and Nigeria—forcing them into three-week-long quarantines (the amount of time it would take for a latent Ebola infection to manifest itself in symptoms).[2] A school board in Maine went so far as to compel a teacher to quarantine herself after she attended a conference in Dallas because it was held ten miles away from a hospital where a man who'd been infected with Ebola in Liberia had been treated.

Any sign of unusual illness in travelers or people perceived to be foreign and thus possibly infected with Ebola similarly elicited elaborate containment and avoidance measures. A frightened flight crew locked a passenger who vomited midflight between Dallas and Chicago in the plane's bathroom, for fear the passenger was suffering from Ebola.[3] A woman who vomited after getting off a shuttle bus at the Pentagon was met by a hazmat team, who isolated her and quarantined the military officers on their way to a Marine Corps ceremony. By November 2014, the Centers for Disease Control felt compelled to assuage consumers with a special note about their upcoming feasts, assuring them that they wouldn't get Ebola from their Thanksgiving turkeys.[4] U.S. politicians inflamed the panic, with one going so far as to warn the CDC that "illegal migrants" from Mexico were carrying Ebola (along with swine flu, dengue, and tuberculosis) over the border into the United States.[5]

The hysteria was not restricted to the United States. In November 2014, the government of Morocco canceled its plans to host the 2015

Africa Cup of Nations soccer match, despite the fact that none of the affected countries had qualified to participate and few visiting fans were expected. American and European travel agencies stopped offering tours anywhere on the continent of Africa.[6] Mexico and Belize refused to allow a cruise ship to dock on their shores because it was carrying a passenger who had handled lab specimens from an Ebola patient in Dallas. The fact that the passenger had not been exposed, was not ill, and had already been quarantined on board did not shake their resolve.[7] In Prague, a student from Ghana shivering on a train platform was carted away by fifteen police officers and an emergency responder wearing a full hazmat suit. It turned out he had a cold. At an airport in Madrid, a crowd of passengers stood aside and watched, frozen in horror, as a Nigerian man collapsed onto the floor and lay there shaking for nearly an hour. He was suffering from a cocaine overdose.[8]

The panicked reactions were so pervasive that they impeded the international effort to control the outbreak in West Africa. While the governments of Guinea, Liberia, and Sierra Leone pled for international help, airlines canceled flights to the affected countries, leaving aid workers stranded. In Australia and Canada, travel to and from West Africa was summarily banned, while elsewhere, draconian quarantine restrictions were put into place for those who visited the affected countries.[9] Hospitals, government agencies, and private citizens stocked up on so many biohazard suits to protect themselves from the hypothetical outbreak that there weren't sufficient stocks left for aid workers en route to fight the actual epidemic in West Africa.[10]

When the bearded man questioned the utility of our fearful reactions to pathogens, he undoubtedly had this seemingly irrational public reaction in mind. What was the point of the spectacle? After all, the risk that Ebola actually posed to the industrialized world was exceedingly low, public-health experts agreed. There were precious few transmission opportunities for a pathogen like Ebola. The virus could spread only if people ingested the bodily fluids of victims in the throes of infection, which was unlikely to happen with any regularity in places where the sick seek treatment in modern hospitals and the handling of corpses is left to professionals. So why were the parents of schoolchildren in Maine and flight crews in Dallas so afraid?

Most commentators chalked it up to ignorance and paranoia. A mocking term to describe the phenomenon, "Ebolanoia," gained popularity on social media. PolitiFact called Americans' fearful exaggerations about Ebola the "Lie of the Year." *The Economist* called it an "Ignorance Epidemic." More Americans had been married to the reality television celebrity Kim Kardashian, commentators pointed out, than had died of Ebola.[11]

But dismissing Ebolanoia as an expression of ignorance missed its larger meaning. Far from being pointless idiocy, the fears that Ebola provoked in the industrialized world revealed something important about prevailing attitudes about pathogens, and how the next pandemic might be received. Fear is a response to the unexpected. Somehow, something about Ebola had violated modern expectations about pathogens and their role in our lives.

Consider how the same societies that panicked about Ebola responded to other new pathogens. Take Lyme disease, which has steadily marched across the country since first emerging in 1975, and which is now diagnosed in three hundred thousand Americans every year. Diagnosis and treatment are tricky. While a prompt course of antibiotics can nip the disease in the bud, because of the difficulty in diagnosing the infection (one out of five people infected with the bacteria doesn't develop the telltale bull's-eye rash that distinguishes the disease, and blood tests are vague and undiscerning), many cases go untreated. Victims then suffer a wide range of long-lasting symptoms, as the infection spreads to the joints, the nervous system, and the heart. Children are especially vulnerable, with boys between the ages of five and nineteen suffering three times more Lyme than adults. And their lives are profoundly disrupted by the pathogen. Children with Lyme suffer symptoms for nearly a year, a CDC study found, and miss more than one hundred days of school on average. A 2011 study found that over 40 percent of children with Lyme suffer suicidal thoughts and 11 percent had made a "suicidal gesture."[12]

And yet, in the epicenter of the epidemic, the disease evokes little more than a collective yawn. New York State accounts for nearly a third of the nation's cases, and Ulster County, home to the leafy campus of the State University of New York in New Paltz, is the eighth-most Lyme-infested county in the nation. I taught a journalism class there during the spring of 2013. Nearly all of the students in the class had been touched by the disease in one way or another. One student's mother had been infected years before and was "never the same" afterward, the student said, plagued by strange, lingering, and mostly untreatable symptoms. Another remembered Christmas gatherings during which young cousins would stumble around, suddenly paralyzed by Lyme. Still another was a survivor of the disease herself. And yet none expressed fear about their own chances of getting infected or professed to taking even the most rudimentary precautions to avoid the bites of ticks. (The CDC recommends applying repellents and wearing insecticide-treated clothing, among other things.) There was zero demand for tick-repellent outdoor clothing at the local outdoor-gear shop. There were no signs about infected ticks along the popular twenty-four-mile hiking trail behind the campus that students and visitors frequented, although ticks would attack by the dozen if one veered even a few feet into the brush. And although nearly all of my students planned to become professional journalists, few saw the Lyme disease epidemic as a newsworthy story.[13]

The emergence of dengue in Florida in 2009 was met with similar indifference. The mosquitoborne disease is known as "breakbone fever" in Asia and Latin America for the intense muscle and joint pain it causes. Most of those who fall ill recover and many who are infected don't fall ill at all. But repeated infections can be deadly. They increase the risk that victims will suffer a more severe form of the disease, including a life-threatening complication called dengue hemorrhagic fever.[14]

Still, the people of Key West scoffed at news of dengue's arrival. Making conclusions about the presence of dengue in a 250,000-strong population on the basis of a sample of a few hundred people—a standard practice that is widely understood to be statistically valid—made no sense, they said. "It's inaccurate," one local said. "If that's the way science works, that just seems really weird to me."[15] The idea that several dozen people suddenly falling ill from a novel virus constituted an "epidemic"—the standard term for such an event—was "very alarmist," another said. "The idea that there's an epidemic here, or that we're on the verge of an epidemic, is just false," a county tourist official added.[16] In a scene reminiscent of the cholera balls of 1832 Paris, a group calling itself Dengue Night Fever pranced through the streets of Key West in the summer of 2010 wearing giant mosquito wings as the virus swirled around them unseen. Undoubtedly, some of the people in Key West pooh-poohing dengue relied in some way on the tourism industry that dengue would threaten. But tourists themselves, *The New York Times* reported, "seemed oblivious" of the ongoing outbreak. One man, with a fresh mosquito bite on his arm, said, "We haven't heard anything about it. We are having a wonderful time."[17] Another visitor had never heard of dengue either, despite being a nursing student and a Florida resident of forty years. "I typically don't wear mosquito repellent," she admitted. "I don't think about it." (Her dengue infection landed her in the emergency room. "It was the worst ten days of my life," she said.)[18]

Of course, every pathogen is different, and its reception is shaped by its specific qualities along with the historical context in which it debuts. Most people in North America and Europe knew, however dimly, that Ebola originated in a distant, exotic place (that is, in a village near the Ebola River, in the Democratic Republic of Congo). That alone may have made it seem inherently more dangerous to Westerners, compared to something named after a leafy Connecticut suburb, like Lyme disease, simply because its birthplace is less familiar. It's also highly virulent. Ebola kills about half its victims, on average. In contrast, Lyme is rarely deadly and dengue hemorrhagic fever kills about 10 percent of its victims.

But none of these specific differences explains why Ebola terrified and other new pathogens did not. The exotic origins of, say, West Nile virus and dengue are obvious, too, and yet neither inspired Ebola-like panic. And if virulence were the main determinant of the fear response, the most terrifying disease would have to be rabies, which kills every one of its victims in a matter of days. But culturally speaking it's more punch line than nightmare. On an episode of the critically acclaimed comedy *The Office*, for example, the show's most absurdly out-of-touch character organizes a "fun run" for rabies awareness (titled "Michael Scott's Dunder Mifflin Scranton Meredith Palmer Memorial Celebrity Rabies Awareness Pro-Am Fun Run Race for the Cure"). The other characters are indifferent, participating by taking taxis and drinking beer and shopping along the way. The joke is that the race organizer ludicrously thinks rabies is a terrifying disease, but the reasonable characters know that it is not.

Ironically, while the seemingly disproportionate alarm triggered by Ebola was roundly condemned, the more dangerous response to new pathogens is probably indifference. One example is the case of our oldest and most resilient pathogen, the one that causes malaria. We've had malaria since we evolved from the apes, and to this day it takes hundreds of thousands of lives every year. And yet it's fully preventable and curable, and has been for centuries. Medical anthropologists have repeatedly found that the reason we still have so much malaria is that many people in malarious societies take few precautions to protect themselves from the disease. They don't sleep under mosquito nets. They don't get diagnosed or treated when they fall ill. Why? Because they think of malaria as a normal problem of life. Malaria persists because it has ceased to inspire fear.

Malaria is an endemic disease in most places where it occurs. Endemic diseases are arguably much worse than epidemic ones. They are harder to get rid of, for the reasons noted above, and more burdensome, too, because they regularly sicken and kill year after year and not just in a single outburst. Cholera has already made the transition from epidemic to endemic status in Haiti. As an endemic disease, it will be a continuous drain on Haitian society for the foreseeable future and a permanent threat to the region. Cases have already appeared in Florida, the Dominican Republic, Cuba, Puerto Rico, Mexico, and the Bahamas.[19]

Dengue is expected to become endemic in Florida, has emerged in Texas, and will likely spread farther north, too, touching millions. Lyme disease is steadily spreading across the United States and drains the economy of hundreds of millions of dollars a year. But once they stop provoking fear, pathogens have secured the golden ticket. They no longer have to mount our defenses because there's little public interest in putting up defenses. The complacency of Floridians and New Yorkers to the new pathogens in their midst is the first step in the cultural and biological process by which itinerant epidemics turn into deeply rooted endemic diseases.

So why do some pathogens provoke yawns while others trigger panic? It may have to do with the way they disrupt or conform to the popular conception of pathogens. This conception is clear from the way we talk about disease. The reigning metaphor, in medicine as in the culture at large, is war. We "attack" illness, we "wage war" on disease, we "arm" ourselves with medicines. "Pandemic disease and war," as *The Economist* put it, "are so similar." But the war we're waging is not against enemies that are elusive or forbidding; on the contrary, they're cast as easy to conquer. Complex, resilient pathogens such as malaria are seen as easily foiled. All it takes is a few dollars. (As one charitable organization put it, "For just $10, you can . . . save a life.") Victory in the war against pathogens is ours, Microsoft cofounder Bill Gates argued in the wake of the 2014 Ebola epidemic. All we have to do is try a little bit harder.[20]

In this popular formulation, we imagine ourselves the victors of a winnable war against pathogens. That may be why the most frightening pathogens are those that seem invulnerable to our armaments, even when they aren't actually threatening to us, like Ebola. The virus loomed over our microbial war machine. During the months of Ebolanoia, there was no vaccine to prevent Ebola, no treatment to cure it. It seemed that even sophisticated Western palliative care—round-the-clock nursing, ventilators, and the like—would not make much difference in the course of the infection. Since Ebola's untamable nature was at the root of the panic, it didn't matter that Ebola is easily and simply avoided. The very fact of its existence disturbed. Ebola was the red six of spades, a painted clown in a darkened cellar: unexpected, unfathomable, terrifying.

This also explains why Lyme, dengue, and rabies, despite being more threatening and burdensome, are considered less scary. Our chemicals can, in theory, vanquish all three. As my students told me, getting a tick bite in Lyme disease country is no big deal: you just pop a course of the antibiotic doxycycline. Both Lyme and dengue are carried by insects, which are vulnerable to a range of lethal and widely available insecticides. And the vaccine against rabies is 100 percent effective. It doesn't seem to matter that we may not actually have these pathogens under control: the existence of a weapon that can be used against them provides the comforting illusion of mastery.

Besides shaping our expectations about the risks that pathogens pose to us, casting certain microbes as enemies obscures the fluidity of their disease-causing capacities and our own complicity in nurturing them. Pathogens become something separate, with fundamentally opposed interests that are unresponsive to us. They are simply our nemeses. And yet, the virulence of even a pandemic-causing pathogen like *Vibrio cholerae* depends entirely on its context. In the body, it's a pathogen; floating in some warm estuary, it's a productive member of a harmonious ecosystem. And much of its transformation from harmless microbe to virulent pathogen has to do with our own activities. We turned it into an enemy ourselves.

The simplistic enemy-victor dichotomy with which we often approach pathogens can't capture this complexity. The result is reactions to pathogens that range from futile paroxysms of fear to lethal apathy.

What we need instead is sustained engagement, with both the formidable threat that pathogens pose and our own critical role in shaping it. That is to say, we need to transcend the simplistic enemy-victor dichotomy and develop a new way of thinking about microbes and our role in the microbial world.

It's already started to happen. The idea that there are "good germs," which are beneficial to human health, and not only "bad germs," has filtered into public consciousness. New ideas about the nature of human health as signifying more than brute victory over pathogenic enemies have started to gain ground, too. The One Health movement, spearheaded by organizations such as Peter Daszak's EcoHealth Alliance, argues that human health is linked to the health of wildlife, livestock, and the ecosystem. In 2007, both the American Medical Association and the American Veterinary Medical Association passed resolutions recognizing the concept, and scientists from the Institute of Medicine, the Centers for Disease Control, and the WHO have signed on. Let's hope that these new ideas about disease will lead to a more honest reckoning with pathogens and the risk they pose. Ultimately, preventing pandemics will require reorganizing human activities that aggravate them. Recognizing the responsive dynamism of the microbial world and our own connection to it is a critical first step.

Of course, reimagining our role in the microbial world—which is to say, reimagining our place in nature—is hardly a quick fix to the menace of pandemics. It's more likely a decades-long, generational project. In the meantime, more immediate measures of pandemic protection will be required.

⊙

If we can't prevent pandemics altogether, the next best thing is to detect them as quickly as possible.

This will require strengthening and expanding the present system of disease surveillance, which is inadequate in several ways. For one, the system is slow and passive. It's triggered only when pathogens make their presence obvious by causing outbreaks. In the United States, the Centers for Disease Control maintain a continually amended list of eighty or so infectious diseases, from syphilis to yellow fever. If a clinician happens upon a patient with one of these "notifiable" diseases, he or she is supposed to inform state-level public-health authorities, who pass on the information to national public-health authorities.[21] If the outbreak has the potential of crossing borders, national authorities must report it to the WHO within twenty-four hours, according to the agency's 2007 "International Health Regulations."[22]

Even when this system works, it's not fast enough. By the time the alarm is triggered, pathogens have already adapted to the human body and cases have already started to grow exponentially. The containment effort required to contain the pathogen will necessarily be large and urgent.

By the time the containment effort began in West Africa in 2014, Ebola had been brewing for months (and possibly much longer than that) in remote forest villages in Guinea. Each victim had infected his or her contacts, and those contacts in turn had infected their contacts, and so on for several iterations, each wave of new infections exponentially larger than the last. Disrupting transmission could have been straightforward—each contact has to be tracked and isolated for Ebola's three-week incubation period—but Ebola had ignited so many concurrent chains of transmission that identifying and isolating all of their potentially exposed contacts had become impossible.[23] By mid-September, when the United States announced plans to send its military to build Ebola treatment units in Liberia, the epidemic there had already peaked. Ultimately, the units they built treated a grand total of twenty-eight patients. Nine of the eleven facilities failed to treat a single one.[24]

Extravagant and only partially effective measures to contain other new pathogens are similarly the result of tardiness. Since H5N1 was not stamped out when it first emerged in the late 1990s, it now regularly plagues poultry flocks around the world. In Hong Kong, authorities slaughter every unsold chicken in their markets nightly in an attempt to contain it.[25] Because the SARS virus was not noticed until it had spread into south China's massive wet markets and started sickening people, containing it required muscular quarantines and travel restrictions that bled Asia's tourism industry of more than $25 billion.[26] Because dengue, West Nile virus, and other vectorborne diseases were not checked before they established footholds across the United States, expensive and controversial campaigns of aerial spraying of insecticides are now de rigueur in many U.S. cities.[27] Even when controlling a disease outbreak requires only the cheapest, easiest fixes—like rehydration therapy for cholera—waiting too long makes it a lot harder. The cholera outbreak in Haiti grew so fast that Doctors Without Borders tapped out the entire global supply of intravenous rehydration fluids.[28] The mismatch between the way epidemics expand and the rollout of even the most well-coordinated containment effort is inevitable: epidemics grow exponentially while our ability to respond proceeds linearly, at best.

The problem is not just that the present surveillance system is slow and passive. It's also riddled with holes. The system gets activated only when a person infected with a notifiable disease arrives at a doctor's office. But that's a reliable switch only if clinicians are trained to detect and report new diseases and their services are readily available across the globe. Neither is the case. People often fail to visit doctors when they're ill. For many, it's too expensive. For others, it's too much trouble. And even when they do anyway, doctors often don't bother diagnosing strange conditions, or reporting them either. I saw this in action myself. A few summers ago, I'd visited my doctor during a debilitating week-long bout of watery diarrhea and vomiting. Rita Colwell speculated that I'd contracted some kind of vibrio infection. But my doctor responded like many probably do when faced with a patient with a strange but likely self-limiting illness they can't easily treat. He didn't order a lab analysis or notify any authorities, though vibrio infections are "notifiable." He shrugged his shoulders. "Probably just some bug," he said, and sent me on my way. Anyone in a similar situation who had been an early victim of a new pathogen would have strolled through the disease surveillance system unnoticed.

There are gaps where no one is watching at all. As of this writing, truckloads of foods and armies of insects carry disease across borders, for the most part free of scrutiny. Hardly anyone tracks the spread of invasive disease-carrying vectors, such as the Asian tiger mosquito, which first arrived in the United States in the mid-1980s. Entomologists proposed containing it before it spread but failed to garner sufficient interest.

Today, the mosquito carries dengue and other diseases, including new mutant strains of a virus called chikungunya, which established itself in the Americas in 2013.[29]

In many countries, even the most rudimentary surveillance is scarce. As of 2013, only 80 of the WHO's 193 member nations had the surveillance capacity to fulfill their WHO-mandated requirements. Antibiotic-resistant pathogens such as NDM-1 are being found basically by accident. There's no national surveillance to track the bug in India. In most avian-flu-affected countries, livestock are not monitored for signs of the virus. Nor are humans, for that matter.[30]

⊙

Fixing the present surveillance system is no small task: it will require easy and affordable health care for people everywhere. A network of clinics, staffed with health-care workers trained to recognize and report new pathogens, could do the trick. At the same time, the surveillance system needs to be significantly expanded. Instead of relying solely on sick people visiting doctors, we can actively search for signs of incipient pandemics.

Obviously it's not possible to surveil every microbe with pandemic potential. Nor is it possible to focus in on a few, for there's no way to know which microbe will cause the next pandemic. But the probability of new pandemic-causing pathogens emerging is not uniform across the globe. There are certain "hot spots" where it's most likely to happen, places where wild habitat is being invaded in new and accelerated ways, crowded slums are booming, factory farms are growing, and air connections are expanding. By actively surveilling these hot spots—and the "sentinel" populations that interact with them—we can zero in on the places most likely to hatch new pandemic-causing pathogens. This kind of active surveillance is already under way. Scientists from the University of Hong Kong, for example, collect hundreds of samples of pig and bird feces monthly from wholesale markets, wildlife preserves, pet shops, and slaughterhouses across Hong Kong. At the blindingly white labs of the Li Ka Shing Institute of Health Sciences, scientists pore over the samples for early signs of pathogens with pandemic potential.[31] USAID's Emerging Pandemic Threats program, founded in 2010, coordinates active surveillance programs in hot spots such as the Congo basin in East and Central Africa, the Mekong region in Southeast Asia, the Amazon in South America, and the Gangetic plain in South Asia.[32] The International Society of Travel Medicine's GeoSentinel program collects information about travelers, who act like canaries in the coal mine for emerging disease, from more than two hundred travel and tropical disease clinics.[33]

A handful of organizations conduct active surveillance for signs of new pathogens in the general population, too. A few U.S. states comb through a continuous stream of data on the "chief complaint" of patients arriving in local emergency rooms, along with data on sales of thermometers and antivirals from pharmacies, for signals that might indicate an impending outbreak. Organizations such as HealthMap and Ascel Bio analyze social media and other online sources to do the same.[34]

These new active surveillance projects have already proved that they can identify outbreaks faster than the traditional passive surveillance system. HealthMap detected the 2014 Ebola outbreak in West Africa nine days before the World Health Organization broke the news. Ascel Bio's James Wilson detected the outbreak of cholera in Haiti weeks before official reports appeared. Active surveillance of hot spots has even fingered new spillover pathogens before they've infected humans at all. In 2012, an active surveillance program founded by the Stanford University virologist Nathan Wolfe uncovered Bas-Congo virus, a novel virus that can cause Ebola-like hemorrhagic fever, in the Democratic Republic of Congo. Earlier efforts by Wolfe, surveilling blood samples collected by bushmeat hunters and wet-market sellers, uncovered several other new microbes, including one called simian foamy virus and another called simian T-lymphotropic virus, both of which have started crossing over into humans.[35]

The risk of epidemics could even be predicted the way meteorologists predict the risk of storms. Weather forecasts and satellite data on chlorophyll signatures could help predict outbreaks of malaria, tickborne disease, and cholera. Thanks to the plummeting price of genetic sequencing, the genomes of the microbes around us—on our computers and toilet handles, and in sewage—can be rapidly and cheaply identified, as the computational biologist Eric Schadt has done, to create microbial maps. Those maps could help scientists pinpoint microbial signatures that precede outbreaks, too.[36]

What emerges from these disparate projects, stitched together alongside beefed-up surveillance based on traditional methods, is a kind of global immune system. It could detect pandemic-worthy pathogens before they hop on flights and get swept up in population movements, pinpointing the next HIV, the next cholera, and the next Ebola before they start to spread.[37] For some enthusiasts, such a system would allow societies to continue doing all the things that make pandemics more likely without suffering the consequences. "I believe you can have your cake and eat it too," the emerging-disease expert Peter Daszak says. "It's not like you have to stop eating meat. Or stop eating spinach," he says, or stop flying in airplanes, or eating food from across the globe, or any of the novel modern behaviors that enhance pathogens' ability to spread. "You can do that. But you can also understand there's a risk. [Then] you have to insure against that risk," he says, by helping to pay for a global system of disease surveillance.[38] A tax on air travel of 1 percent could cover the bill.[39] A global pandemic insurance fund could disburse the money necessary to respond to these early alarms, the same way that disaster insurance policies pay out when hurricanes and earthquakes strike. The World Bank and the African Union started to discuss establishing such a fund in the spring of 2015.[40]

It's an appealingly technocratic approach. Early notice makes possible all kinds of more efficient containment and mitigation. We could prevent some epidemics and more effectively prepare to withstand others. But even if such a global surveillance system can be built, it will work only if it translates into people actually using the information to do something about it. And as challenging as building an active global surveillance system will be, ensuring that people act will require an even more far-reaching global project.

⊙

The remote fishing village of Belle-Anse, on the southwestern coast of Haiti, is like countless other towns and villages where the world's poor live, ostensibly connected to the global economy and yet brutally isolated. It's about fifty miles from Port-au-Prince. I made the trip in the summer of 2013. It began in a thirty-year-old Nissan minivan, built to accommodate around eight passengers but carrying nearly twenty that day, including a couple with a small whimpering child and a man with a surprisingly sedate chicken on his lap. The minivan took us over the mountains on steep, narrow, winding roads, discharging us at a dusty lot in the foothills by the coast. But that was just the beginning of the journey. An hour-long motorcycle ride to the shore followed, and then, because the road to Belle-Anse is impassable, another hour zooming across long, wide swells on a fifteen-foot skiff with an outboard motor tied to its stern with a frayed bit of rope.

It took us eight hours to get to Belle-Anse from the capital city. It took cholera about a year. The epidemic enveloped most of the country in 2010 but didn't reach Belle-Anse until 2011. Its arrival was not only predictable: it *had* been predicted, using many of the digitally enhanced techniques that will power the new active global surveillance system. There'd been a huge influx of NGOs onto the island before the epidemic started, due to the earthquake that had occurred earlier in the year. When cholera broke out, they used all the technology at their disposal to track its spread. The epidemiologist Jim Wilson and his team trolled Twitter feeds and handed out their cell numbers to local people across the country. "We would see cholera just marching down the highway," he remembered.[41] Volunteers of all kinds had mapped the entire country, "down to each stray dog," one aid worker said. A Swedish NGO, in collaboration with the local mobile phone company, tracked people's SIM cards in their phones to map their movements so that they could predict where cholera would hit next. As an early trial of the kind of active disease surveillance system that could be extended globally, it worked beautifully. Cholera arrived in Belle-Anse in 2011 as the Twitter charters and the SIM card trackers watched.[42]

But that early detection made no difference in the course of the disease. There was no early containment effort that could have short-circuited the outbreak in Belle-Anse. On the contrary, cholera killed people there at a rate four times higher than the rest of the country. By the time I visited, the sole NGO that had set up a cholera treatment center in the town center had already left, and people who fell ill in the hills above the town were being found dead along the three-mile-long descent into town. The local authorities could do little more than send body bags.[43]

Ironically, this wasn't due to the village's remoteness or to a lack of foreign aid. In fact, it was an aid project and a nascent transportation network that had made the people of Belle-Anse vulnerable to cholera in the first place. Due to a failing water supply system, built in the mid-1980s by the Belgian government, the people of Belle-Anse had very little access to clean fresh water. The system involved a pipe that had been laid across a high, long ridge, which carried fresh water from the hills into the town. The Belgians apparently hadn't considered Haitian geography, climate, or the ability of locals to maintain the system when they installed it. Because of Haiti's eroding hillsides and tropical storms, every year the pipe slipped off the ridge toward the beach. Today, it rests just inches above the turquoise waves, vulnerable to the force of hurricanes and peppered with holes. Locals don't have the right tools to fix the pipe or sufficient resources to get them, so they wrap the holes with cloth bandages secured with elastic. They leak nonetheless. Only a trickle of fresh water arrives in the village as a result, necessitating all the shortcuts that diminish people's ability to avoid filth and its pathogens.[44]

Just as ill-conceived aid had exposed Belle-Anse to pathogens, so had a partially functional, inaccessible transportation system. On one hand, Belle-Anse was viscerally connected to the world of commerce and disease around it. It was because Belle-Anse was connected to the rest of the country that cholera had arrived, after all. On the other hand, while the transportation system was sufficient to bring in the disease, it couldn't bring in sufficient aid or resources to stanch it, and it couldn't carry away to safety those who wanted to leave, either. Three days before we'd arrived in Belle-Anse, some residents desperate to escape had stolen a skiff like the one we rode over in. But unlike us, they didn't have access to life jackets. When the overloaded boat capsized on the crossing, four drowned. On our boat ride back out of Belle-Anse, we saw the swollen body of one of the would-be escapees, a three-year-old girl dressed in pink leggings, bobbing in the sea. There was no room in our boat to take her corpse with us. We cut the motor and silently swayed alongside her, while our skipper made some calls to report the location of her watery grave.

Many simple solutions and quick fixes have been sought to address the lack of governance and poverty that prevented Belle-Anse from protecting itself from cholera. The history of foreign aid is littered with them. But it seems that there are no easy answers. The first step, perhaps, is simply to accept that a sustained, multifaceted effort will be required.

⊙

I returned home from Haiti just as I had from other trips far afield, struck by the power of pathogens to transform society. Their disruptive force looms over all of us, although only some have felt it so far. And yet even as the next pandemic gains momentum, the identity of the pathogen that will cause it remains as obscure as ever. It could be a jungle pathogen like Ebola. It could be a marine creature like cholera. It could be something else entirely. Somehow we have to live with the fact of not knowing its name.

So I wondered one summer evening, en route to the Chesapeake Bay to escape the heat of the city. The opaque, brackish water was as warm as a bath and rich with life. There were schools of striped bass and blue-fish, beds of seagrass with crabs tangled in their waving blades, and floating armies of plankton. There were vibrio bacteria in the water, too, I knew, including cholera. It lapped at the curved sides of the fiberglass boat onto which we'd clambered.

Still, while the water was warm, the air was warmer, and it felt lovely to slip off the side of the boat into its fluid embrace. The bay is not deep, but it took a while before I reached the soft silt blanketing its floor. Above, the cholera-rich waters displaced by my dive whirled around, glittering darkly.

GLOSSARY

antibiotic. A compound that either kills or slows the growth of bacteria, used in the treatment of bacterial infections.

antibodies. Proteins produced by the immune system to identify and neutralize pathogens.

bacteria (singular: **bacterium**). Microscopic single-celled organisms.

bacteriophage. A virus that infects bacteria.

basic reproductive number. The average number of susceptible people who are infected by a single infected person, in the absence of outside intervention.

Batrachochytrium dendrobatidis. The fungal pathogen responsible for widespread declines in amphibians around the world; also known as amphibian chytrid fungus.

Borrelia burgdorferi. The tickborne bacterium that causes Lyme disease.

contagion. An infectious disease that spreads through direct or indirect contact.

copepods. A group of tiny marine crustaceans that are often colonized by vibrio bacteria, including *Vibrio cholerae.*

coronaviruses. A genus that includes the virus that causes SARS and MERS.

E. coli (Escherichia coli). A bacterium found in the gut of warm-blooded animals.

emerging diseases. New diseases that have caused increasing numbers of cases in recent decades and are poised to continue to do so.

epidemic. An unusual increase in the occurrence of a disease in a particular locality, often caused by contagion.

excreta. Matter that is excreted from the body, such as urine, feces, and saliva.

gene. A segment of DNA that is the basic unit of heredity.

H1N1. A subtype of influenza that caused the 1918 influenza pandemic and the 2009 "swine flu" pandemic.

H5N1. Also known as "bird flu," a subtype of avian influenza that first emerged in 1996 and is highly virulent in humans.

hemorrhagic fever. Fever caused by a viral infection, accompanied by heightened susceptibility to bleeding.

highly pathogenic avian influenza. Strains of avian influenza that are highly pathogenic, such as H5N1.

horizontal gene transfer. A method of transferring genes laterally, common in single-celled organisms.

MERS (Middle East respiratory syndrome). An emerging infectious disease first reported in 2012, caused by a coronavirus.

microbe. Any organism too small to be seen with the naked eye.

monkeypox. A virus, related to smallpox, that lives in rodents and causes a disease clinically indistinguishable from smallpox in humans.

MRSA (methicillin-resistant *Staphylococcus aureus*). A bacterium that causes a range of difficult-to-treat infections in humans.

NDM-1 (New Delhi metallo-beta-lactamase 1). A plasmid that enables bacteria to resist fourteen classes of antibiotics.

Nipah virus. A virus of bats first reported in humans in 1999.

outbreak. A sudden increase in or eruption of disease.

pandemic. An infectious disease that spreads out of a particular locality to infect populations across regions or continents.

pathogen. Any disease-causing organism.

Phythophthora infestans. Fungal pathogen that causes potato blight and was the causative agent of the 1845 Irish potato famine.

plankton. A diverse range of marine organisms that float in the water column and cannot swim.

plasmid. A fragment of DNA within cells that can spread and replicate independently.

Plasmodium falciparum. A parasitic pathogen that causes a deadly strain of malaria in humans.

Pseudogymnoascus destructans. The fungal pathogen that causes white-nose syndrome in bats.

quarantine. Isolation to prevent the spread of infectious disease.

reemerging disease. An old disease causing increasing numbers of cases or spreading into new areas.

SARS (severe acute respiratory syndrome). An infectious disease first reported in 2003, which is caused by a novel coronavirus.

spillover. The process by which a microbe in one species starts infecting a different species.

STEC (Shiga toxin–producing *E. coli*). A strain of *E. coli* found in some cattle, which causes virulent disease in humans.

vectorborne pathogens. Pathogens carried from one host to another via a vector, such as an insect.

vibrio. Any of the bacteria from the genus *Vibrio.*

vibrio bacteria. A genus of marine bacteria that includes both pathogenic and nonpathogenic species.

Vibrio cholerae. The bacterial pathogen that is the causative agent of cholera.

virion. A single virus particle.

virulence. A measure of a pathogen's ability to cause disease.

virus. Microscopic agents that replicate inside the living cells of other organisms.

zoonosis. An infectious disease of animals that can infect humans.

zooplankton. Animal-like plankton.

NOTES

PREFACE TO THE 2020 PAPERBACK EDITION
1. https://twitter.com/amymaxmen/status/1233420175317168128.
2. *CBC Sports*. (Tuesday, March 24, 2020). COC's early action helped guide IOC toward decision it ultimately wanted. *CBC News*. Retrieved from https://advance -lexis-com.proxy-tu.researchport.umd.edu/api/document?collection=news&id=ur n:contentItem:5YHC-TN31-JCSH-34J6-00000-00&context=1516831.
3. https://www.theatlantic.com/politics/archive/2020/04/world-health-organization -blame-pandemic-coronavirus/609820/.
4. https://www.reuters.com/article/us-health-coronavirus-usa-hart-island/new-york -city-hires-laborers-to-bury-dead-in-hart-island-potters-field-amid-coronavirus -surge-idUSKCN21R398.
5. https://www.washingtonpost.com/graphics/2020/world/iran-coronavirus-outbreak -graves/; https://time.com/5811222/wuhan-coronavirus-death-toll/.
6. https://www.propublica.org/article/this-coronavirus-is-unlike-anything-in-our -lifetime-and-we-have-to-stop-comparing-it-to-the-flu.
7. https://theweek.com/speedreads/903635/trump-insisted-nobody-ever-thought -pandemic-like-covid19-administration-did-last-year.
8. https://www.nytimes.com/2020/04/06/world/europe/coronavirus-terrorism-threat -response.html.
9. https://www.cnn.com/2020/04/13/us/coronavirus-made-in-lab-poll-trnd/index.html ?fbclid=IwAR0hRJESJ1q3J7uPB3CJPfes9z6q6Y2O_MVqNwXi5Dc4cJ2OHfr MqMfK0ko.
10. https://www.latimes.com/entertainment-arts/story/2020-04-13/coronavirus-bill -gates-ellen-degeneres-ted-talk.
11. https://www.rollingstone.com/tv/tv-news/bill-gates-coronavirus-pandemic-ted -talk-daily-show-trevor-noah-977869/.
12. https://foreignpolicy.com/2020/03/20/world-order-after-coroanvirus-pandemic/.

13. https://www.politico.com/news/magazine/2020/03/19/coronavirus-effect-economy
-life-society-analysis-covid-135579.
14. https://www.washingtonpost.com/outlook/2020/03/20/what-will-have-changed
-forever-after-coronavirus-abates/?arc404=true.
15. https://www.eventbrite.com/e/the-pandemic-is-a-portal-a-conversation-with
-arundhati-roy-tickets-102343167168.
16. https://journals.lww.com/cmj/Abstract/publishahead/Identification_of_a_novel
_coronavirus_causing.99423.aspx.
17. https://www.theatlantic.com/ideas/archive/2020/03/china-trolling-world-and
-avoiding-blame/608332/.
18. https://www.cnbc.com/2020/03/17/coronavirus-france-president-macron-warns
-we-are-at-war.html.
19. http://www.chinatoday.com.cn/ctenglish/2018/tpxw/202002/t20200211_800192462
.html.
20. https://nationalinterest.org/feature/why-trump-cant-go-war-against-coronavirus
-140777.
21. Ibid.
22. http://jamaica-star.com/article/news/20200319/man-beaten-after-sneezing-bus.
23. https://www.dailymail.co.uk/news/article-7944835/Bystanders-refused-CPR-man
-heart-attack-died-fears-coronavirus.html.
24. https://www.ft.com/content/1eeedb71-d9dc-4b13-9b45-fcb7898ae9e1.
25. https://www.miamiherald.com/news/coronavirus/article241967281.html.
26. https://www.nytimes.com/2020/04/16/health/WHO-Trump-coronavirus.html.

INTRODUCTION: CHOLERA'S CHILD

1. Rita Colwell, "Global Climate and Infectious Disease: The Cholera Paradigm,"
Science 274, no. 5295 (1996): 2025–31.
2. M. Burnet, *Natural History of Infectious Disease* (Cambridge: Cambridge University
Press, 1962), cited in Gerald B. Pier, "On the Greatly Exaggerated Reports of the
Death of Infectious Diseases," *Clin Infectious Diseases* 47, no. 8 (2008): 1113–14.
3. Madeline Drexler, *Secret Agents: The Menace of Emerging Infections* (Washington, DC: Joseph Henry Press, 2002), 6.
4. Kristin Harper and George Armelagos, "The Changing Disease-Scape in the
Third Epidemiological Transition," *International Journal of Environmental Research and Public Health* 7, no. 2 (2010): 675–97.
5. Peter Washer, *Emerging Infectious Diseases and Society* (New York: Palgrave Macmillan, 2010), 47.
6. Kate E. Jones et al., "Global Trends in Emerging Infectious Diseases," *Nature* 451,
no. 7181 (2008): 990–93.
7. Stephen Morse, plenary address, International Society for Disease Surveillance,
Atlanta, GA, Dec. 7–8, 2011.
8. Burnet, *Natural History of Infectious Disease.*
9. Jones, "Global Trends in Emerging Infectious Diseases."
10. Paul W. Ewald and Gregory M. Cochran, "*Chlamydia pneumoniae* and Cardiovas-
cular Disease: An Evolutionary Perspective on Infectious Causation and Antibiotic
Treatment," *The Journal of Infectious Diseases* 181, supp. 3 (2000): S394–S401.

11. Brad Spellberg, "Antimicrobial Resistance: Policy Recommendations to Save Lives," International Conference on Emerging Infectious Diseases, Atlanta, GA, March 13, 2012.

12. Drexler, *Secret Agents*, 7.

13. This survey was cited by epidemiologist Larry Brilliant in a 2006 TED Talk. Larry Brilliant, "My wish: Help me stop pandemics," TED, February 2006.

14. Fatimah S. Dawood et al., "Estimated Global Mortality Associated with the First 12 Months of 2009 Pandemic Influenza A H1N1 Virus Circulation: A Modelling Study," *The Lancet Infectious Diseases* 12, no. 9 (2012): 687–95.

15. Ronald Barrett et al., "Emerging and Re-emerging Infectious Diseases: The Third Epidemiologic Transition," *Annual Review of Anthropology* 27 (1998): 247–71.

16. World Health Organization, "Ebola Response Roadmap—Situation Report," May 6, 2015; "UN Says Nearly $1.26 Billion Needed to Fight Ebola Outbreak," *The Straits Times*, Sept. 16, 2014; Daniel Schwartz, "Worst-ever Ebola Outbreak Getting Even Worse: By the Numbers," *CBCnews*, CBC/Radio-Canada, Sept. 16, 2014; Denise Grady, "U.S. Scientists See Long Fight Against Ebola," *The New York Times*, Sept. 12, 2014.

17. CDC, "U.S. Multi-State Measles Outbreak 2014–2015," Feb. 12, 2015; CDC, "Notes from the Field: Measles Outbreak—Indiana, June–July 2011," *MMWR*, Sept. 2, 2011.

18. Maryn McKenna, *Superbug: The Fatal Menace of MRSA* (New York: Free Press, 2010), 34; Andrew Pollack, "Looking for a Superbug Killer," *The New York Times*, Nov. 6, 2010.

19. N. Cimolai, "MRSA and the Environment: Implications for Comprehensive Control Measures," *European Journal of Clinical Microbiology & Infectious Diseases* 27, no. 7 (2008): 481–93.

20. Interview with Rita Colwell, Sept. 23, 2011.

21. Dawood, "Estimated Global Mortality"; Cecile Viboud et al., "Preliminary Estimates of Mortality and Years of Life Lost Associated with the 2009 A/H1N1 Pandemic in the US and Comparison with Past Influenza Seasons," *PLoS Currents* 2 (March 2010).

1. THE JUMP

1. Rachel M. Wasser and Priscilla Bei Jiao, "Understanding the Motivations: The First Step Toward Influencing China's Unsustainable Wildlife Consumption," TRAFFIC East Asia, Jan. 2010.

2. Y. Guan, et al., "Isolation and Characterization of Viruses Related to the SARS Coronavirus from Animals in Southern China," *Science* 302, no. 5643 (2003): 276–78.

3. Tomoki Yoshikawa et al., "Severe Acute Respiratory Syndrome (SARS) Coronavirus-Induced Lung Epithelial Cytokines Exacerbate SARS Pathogenesis by Modulating Intrinsic Functions of Monocyte-Derived Macrophages and Dendritic Cells," *Journal of Virology* 83, no. 7 (April 2009): 3039–48.

4. Guillaume Constantin de Magny et al., "Role of Zooplankton Diversity in *Vibrio cholerae* Population Dynamics and in the Incidence of Cholera in the Bangladesh Sundarbans," *Applied and Environmental Microbiology* 77, no. 17 (Sept. 2011).

5. Arthur G. Humes, "How Many Copepods?" *Hydrobiologia* 292/293, no. 1–7 (1994).

6. C. Yu et al., "Chitin Utilization by Marine Bacteria. A Physiological Function for Bacterial Adhesion to Immobilized Carbohydrates," *The Journal of Biological Chemistry* 266 (1991): 24260–67; Carla Pruzzo, Luigi Vezzulli, and Rita R. Colwell, "Global Impact of *Vibrio cholerae* Interactions with Chitin," *Environmental Microbiology* 10, no. 6 (2008): 1400–10.

7. Brij Gopal and Malavika Chauhan, "Biodiversity and Its Conservation in the Sundarban Mangrove Ecosystem," *Aquatic Sciences* 68, no. 3 (Sept. 4, 2006): 338–54; Ranjan Chakrabarti, "Local People and the Global Tiger: An Environmental History of the Sundarbans," *Global Environment* 3 (2009): 72–95; J. F. Richards and E. P. Flint, "Long-Term Transformations in the Sundarbans Wetlands Forests of Bengal," *Agriculture and Human Values* 7, no. 2 (1990): 17–33; R. M. Eaton, "Human Settlement and Colonization in the Sundarbans, 1200–1750," *Agriculture and Human Values* 7, no. 2 (1990): 6–16.

8. Paul Greenough, "Hunter's Drowned Land: Wonderland Science in the Victorian Sundarbans," in John Seidensticker et al., eds., *The Commons in South Asia: Societal Pressures and Environmental Integrity in the Sundarbans of Bangladesh* (Washington, DC: Smithsonian Institution, International Center, workshop, Nov. 20–21, 1987).

9. Eaton, "Human Settlement and Colonization in the Sundarbans"; Richards and Flint, "Long-Term Transformations in the Sundarbans Wetlands Forests of Bengal."

10. Rita R. Colwell, "Oceans and Human Health: A Symbiotic Relationship Between People and the Sea," American Society of Limnology and Oceanography and the Oceanographic Society, Ocean Research Conference, Honolulu, Feb. 16, 2004.

11. The filament is called the toxin coregulated pilus or TCP. Juliana Li et al., "*Vibrio cholerae* Toxin-Coregulated Pilus Structure Analyzed by Hydrogen/Deuterium Exchange Mass Spectrometry," *Structure* 16, no. 1 (2008): 137–48.

12. Kerry Brandis, "Fluid Physiology," Anaesthesia Education, www.anaesthsiaMCQ .com; Paul W. Ewald, *Evolution of Infectious Disease* (New York: Oxford University Press, 1994), 25.

13. Zindoga Mukandavire, David L. Smith, and J. Glenn Morris, Jr., "Cholera in Haiti: Reproductive Numbers and Vaccination Coverage Estimates," *Scientific Reports* 3 (2013).

14. Ewald, *Evolution of Infectious Disease*, 25.

15. Dhiman Barua and William B. Greenough, eds., *Cholera* (New York: Plenum Publishing, 1992).

16. Jones, "Global Trends in Emerging Infectious Diseases."

17. N. D. Wolfe, C. P. Dunavan, and J. Diamond, "Origins of Major Human Infectious Diseases, *Nature* 447, no. 7142 (2007): 279–83; Jared Diamond, *Guns, Germs, and Steel: The Fates of Human Societies* (New York: Norton, 1997), 207.

18. Interview with Peter Daszak, Oct. 28, 2011.

19. Lee Berger et al., "Chytridiomycosis Causes Amphibian Mortality Associated with Population Declines in the Rain Forests of Australia and Central America," *Proceedings of the National Academy of Sciences* 95, no. 15 (1998): 9031–36.

20. Mark Woolhouse and Eleanor Gaunt, "Ecological Origins of Novel Human Pathogens," *Critical Reviews in Microbiology* 33, no. 4 (2007): 231–42.

21. Keith Graham, "Atlanta and the World," *The Atlanta Journal-Constitution*, Nov. 12, 1998.

22. "Restoring the Battered and Broken Environment of Liberia: One of the Keys to a New and Sustainable Future," United Nations Environment Programme, Feb. 13, 2004.

23. "Sub-regional Overview," *Africa Environment Outlook* 2, United Nations Environment Programme, 2006; "Deforestation in Guinea's Parrot's Beak Area: Image of the Day," NASA, http://earthobservatory.nasa.gov/IOTD/view.php?id=6450.

24. P. M. Gorresen and M. R. Willig, "Landscape Responses of Bats to Habitat Fragmentation in Atlantic Forest of Paraguay," *Journal of Mammalogy* 85 (2004): 688–97.

25. Charles H. Calisher et al., "Bats: Important Reservoir Hosts of Emerging Viruses," *Clinical Microbiology Reviews* 19, no. 3 (2006): 531–45; Andrew P. Dobson, "What Links Bats to Emerging Infectious Diseases?" *Science* 310, no. 5748 (2005): 628–29; Dennis Normile et al., "Researchers Tie Deadly SARS Virus to Bats," *Science* 309, no. 5744 (2005): 2154–55.

26. Dobson, "What Links Bats to Emerging Infectious Diseases?"; Sonia Shah, "The Spread of New Diseases: The Climate Connection," *Yale Environment* 360 (Oct. 15, 2009).

27. Randal J. Schoepp et al., "Undiagnosed Acute Viral Febrile Illnesses, Sierra Leone," *Emerging Infectious Diseases*, July 2014.

28. Pierre Becquart et al., "High Prevalence of Both Humoral and Cellular Immunity to Zaire Ebolavirus Among Rural Populations in Gabon," *PLoS ONE* 5, no. 2 (2010): e9126.

29. Sudarsan Raghavan, "'We Are Suffering': Impoverished Guinea Offers Refugees No Ease," *San Jose Mercury News*, Feb. 25, 2001.

30. Daniel G. Bausch, "Outbreak of Ebola Virus Disease in Guinea: Where Ecology Meets Economy," *PLoS Neglected Tropical Diseases*, July 31, 2014; Sylvain Baize et al., "Emergence of Zaire Ebola Virus Disease in Guinea—Preliminary Report," *The New England Journal of Medicine*, April 16, 2014.

31. "Ebola in West Africa," *The Lancet Infectious Diseases* 14, no. 9 (Sept. 2014).

32. C. L. Althaus, "Estimating the Reproduction Number of Ebola Virus (EBOV) During the 2014 Outbreak in West Africa," *PLoS Currents Outbreaks*, Sept. 2, 2014.

33. "UN Announces Mission to Combat Ebola, Declares Outbreak 'Threat to Peace and Security,'" UN News Centre, Sept. 18, 2014.

34. Denise Grady, "Ebola Cases Could Reach 1.4 Million Within Four Months, CDC Estimates," *The New York Times*, Sept. 23, 2014.

35. Sadie J. Ryan and Peter D. Walsh, "Consequences of Non-Intervention for Infectious Disease in African Great Apes," *PLoS ONE* 6, no. 12 (2011): e29030.

36. Interview with Anne Rimoin, Sept. 27, 2011.

37. A. W. Rimoin et al., "Major Increase in Human Monkeypox Incidence 30 Years After Smallpox Vaccination Campaigns Cease in the Democratic Republic of Congo," *Proceedings of the National Academy of Sciences of the United States of America* 107, no. 37 (2010): 16262–67.

38. D. S. Wilkie and J. F. Carpenter, "Bushmeat Hunting in the Congo Basin: An As-

sessment of Impacts and Options for Mitigation," *Biodiversity and Conservation* 8, no. 7 (1999): 927–55.

39. Sonia Shah, "Could Monkeypox Take Over Where Smallpox Left Off?" *Scientific American*, March 2013.

40. J. O. Lloyd-Smith, "Quantifying the Risk of Human Monkeypox Emergence in the Aftermath of Smallpox Eradication," Epidemics: Third International Conference on Infectious Disease Dynamics, Boston, Nov. 30, 2011.

41. Dennis Normile, "Up Close and Personal with SARS," *Science* 300, no. 5621 (2003): 886–87.

42. "The Dog That's Just Dyeing to Be a Tiger: How Chinese Owners Turn Their Pets into Exotic Wildlife in New Craze," *Daily Mail Online*, June 9, 2010; John Knight, ed., *Wildlife in Asia: Cultural Perspectives* (New York: Routledge, 2004); S. A. Mainka and J. A. Mills, "Wildlife and Traditional Chinese Medicine: Supply and Demand for Wildlife Species," *Journal of Zoo and Wildlife Medicine* 26, no. 2 (1995): 193–200.

43. Knight, *Wildlife in Asia*.

44. Lauren Swanson, "1.19850+Billion Mouths to Feed: Food Linguistics and Cross-Cultural, Cross-'National' Food Consumption Habits in China," *British Food Journal* 98, no. 6 (1996): 33–44.

45. Anthony Kuhn, "A Chinese Imperial Feast a Year in the Eating," NPR, Jan. 9, 2010.

46. Eoin Gleeson, "How China Fell in Love with Louis Vuitton," *MoneyWeek*, June 14, 2007.

47. Interview with Jonathan Epstein, Sept. 17, 2009; L. M. Looi et al., "Lessons from the Nipah Virus Outbreak in Malaysia," *The Malaysian Journal of Pathology* 29, no. 2 (2007): 63–67.

48. A. Townsend Peterson et al., "Predictable Ecology and Geography of West Nile Virus Transmission in the Central United States," *Journal of Vector Ecology* 33, no. 2 (2008): 342–52; A. Townsend Peterson et al., "West Nile Virus: A Reemerging Global Pathogen," *Emerging Infectious Diseases* 7, no. 4 (2001): 611–14.

49. Most of these people had "silent" infections; they were not sick. Drexler, *Secret Agents*, 72.

50. A. Marm Kilpatrick, "Globalization, Land Use, and the Invasion of West Nile Virus," *Science*, Oct. 21, 2011; Valerie J. McKenzie and Nicolas E. Goulet, "Bird Community Composition Linked to Human West Nile Virus Cases Along the Colorado Front Range," *EcoHealth*, Dec. 2, 2010.

51. Richard Ostfeld, "Ecological Drivers of Tickborne Diseases in North America," International Conference on Emerging Infectious Diseases, Atlanta, GA, March 13, 2012.

52. "CDC Provides Estimate of Americans Diagnosed with Lyme Disease Each Year," Centers for Disease Control and Prevention, Aug. 19, 2013; Julie T. Joseph et al., "Babesiosis in Lower Hudson Valley, New York, USA," *Emerging Infectious Diseases* 17 (May 26, 2011); Laurie Tarkan, "Once Rare, Infection by Tick Bites Spreads," *The New York Times*, June 20, 2011.

53. Felicia Keesing et al., "Impacts of Biodiversity on the Emergence and Transmission of Infectious Diseases," *Nature* 468 (Dec. 2, 2010): 647–52.

54. Beth Mole, "MRSA: Farming Up Trouble," *Nature*, July 24, 2013.

55. Drexler, *Secret Agents*, 136.

2. LOCOMOTION
 1. "Control of Communicable Diseases, Restrictions on African Rodents, Prairie Dogs and Certain Other Animals," Food and Drug Administration, Federal Register, Sept. 8, 2008.
 2. M. G. Reynolds et al., "A Silent Enzootic of an Orthpoxvirus in Ghana, West Africa: Evidence for Multi-Species Involvement in the Absence of Widespread Human Disease," *The American Journal of Tropical Medicine and Hygiene* 82, no. 4 (April 2010): 746–54.
 3. Interview with Mark Slifka, Boston, Nov. 30, 2011.
 4. Lisa Warnecke et al., "Inoculation of Bats with European *Pseudogymnoascus destructans* Supports the Novel Pathogen Hypothesis for the Origin of White-nose Syndrome," *Proceedings of the National Academy of Sciences* 109, no. 18 (2012): 6999–7003; "White-Nose Syndrome (WNS)," USGS National Wildlife Health Center, www.nwhc.usgs.gov/disease_information/white-nose_syndrome/.
 5. Emily Badger, "We've Been Looking at the Spread of Global Pandemics All Wrong," *The Atlantic*, CityLab, Feb. 25, 2013.
 6. "Threading the Climate Needle: The Agulhas Current System," National Science Foundation, April 27, 2011.
 7. C. Razouls et al., "Diversity and Geographic Distribution of Marine Planktonic Copepods," http://copepodes.obs-banyuls.fr/en.
 8. François Delaporte, *Disease and Civilization: The Cholera in Paris, 1832* (Cambridge, MA: MIT Press, 1986), 40.
 9. Walter Benjamin, "Paris—Capital of the Nineteenth Century," *Perspecta*, 12 (1969).
10. N. P. Willis, *Prose Works* (Philadelphia: Carey and Hart, 1849).
11. Roy Porter, *The Greatest Benefit to Mankind: A Medical History of Humanity* (New York: Norton, 1997), 308–10.
12. Strains of *Vibrio cholerae* may have developed elsewhere in the world as well. There are historical accounts of earlier outbreaks that sound a lot like cholera. Ancient Sanskrit writings from 500 to 400 B.C. describe a cholera-like disease, as do historical accounts from ancient Greece and Rome. By the time Vasco da Gama landed on the Malabar coast of India in 1498, about twenty thousand men had already died of a disease that "struck them sudden-like in the belly, so that some of them died in 8 hours." Thomas Sydenham described a cholera-like disease in Britain in 1669 and Rudyard Kipling described a scourge that destroyed visitors to Africa within twenty-four hours that could have been cholera, too. But it was in the Sundarbans that the first global pandemic began, and scientists believe that there was something particularly transmissible about the cholera vibrios that developed there. See Joan L. Aron and Jonathan A. Patz, eds., *Ecosystem Change and Public Health: A Global Perspective* (Baltimore: Johns Hopkins University Press, 2001), 328; Colwell, "Global Climate and Infectious Disease."
13. Myron Echenberg, *Africa in the Time of Cholera: A History of Pandemics from 1817 to the Present* (New York: Cambridge University Press, 2011), 7.
14. Richard J. Evans, *Death in Hamburg: Society and Politics in the Cholera Years* (New York: Penguin, 2005), 229.
15. Washer, *Emerging Infectious Diseases*, 153.
16. Evans, *Death in Hamburg*, 229.
17. Marc Alexander, "'The Rigid Embrace of the Narrow House': Premature Burial & the Signs of Death," *The Hastings Center Report* 10, no. 3 (June 1980): 25–31.

18. Delaporte, *Disease and Civilization*, 43.
19. Ibid., 27–48; N. P. Willis, "Letter XVIII: Cholera—Universal terror . . ." and "Letter XVI: the cholera—a masque ball—the gay world—mobs—visit to the hotel dieu," *Pencillings by the Way* (New York: Morris & Willis, 1844).
20. Delaporte, *Disease and Civilization*, 40, 43.
21. Edward P. Richards, Katharine C. Rathbun, and Jay Gold, "The Smallpox Vaccination Campaign of 2003: Why Did It Fail and What Are the Lessons for Bioterrorism Preparedness?" *Louisiana Law Review* 64 (2004).
22. Willis, *Prose Works*.
23. Bank of the Manhattan Company, "Ships and Shipping of Old New York: A Brief Account of the Interesting Phases of the Commerce of New York from the Foundation of the City to the Beginning of the Civil War" (New York, 1915), 39.
24. Even first-class passengers had to endure discomforts. Private sleeping cabins were cold, poorly ventilated, and dim, with beds consisting of just thin sacking over a board, hollowed out in the middle to prevent sleeping passengers from rolling off in heavy seas. Stephen Fox, *The Ocean Railway: Isambard Kingdom Brunel, Samuel Cunard and the Revolutionary World of the Great Atlantic Steamships* (New York: Harper, 2003), 7–14; "On the Water," *Maritime Nation, 1800–1850: Enterprise on the Water*, Smithsonian National Museum of American History, http://americanhistory.si.edu/onthewater/exhibition/2_3.html.
25. Echenberg, *Africa in the Time of Cholera*, 61.
26. J. S. Chambers, *The Conquest of Cholera: America's Greatest Scourge* (New York: Macmillan, 1938), 298.
27. J. T. Carlton, "The Scale and Ecological Consequences of Biological Invasions in the World's Oceans," in Odd Terje Sandlund et al., eds., *Invasive Species and Biodiversity Management* (Boston: Kluwer Academic, 1999); Mike McCarthy, "The Iron Hull: A Brief History of Iron Shipbuilding," *Iron Ships & Steam Shipwrecks: Papers from the First Australian Seminar on the Management of Iron Vessels & Steam Shipwrecks* (Fremantle: Western Australian Maritime Museum, 1985).
28. Rita R. Colwell et al., "Global Spread of Microorganisms by Ships," *Nature* 408, no. 6808 (2000): 49.
29. Chambers, *The Conquest of Cholera*, 201; Carol Sheriff, *The Artificial River: The Erie Canal and the Paradox of Progress, 1817–1862* (New York: Hill & Wang, 1996), 15–17.
30. Steven Solomon, *Water: The Epic Struggle for Wealth, Power, and Civilization* (New York: Harper, 2010), 289.
31. Ashleigh R. Tuite, Christina H. Chan, and David N. Fisman, "Cholera, Canals, and Contagion: Rediscovering Dr Beck's Report," *Journal of Public Health Policy* 32, no. 3 (Aug. 2011); Maximilian, Prince of Wied, "Early Western Travels, vol. 22: Part I of Maximilian, Prince of Weid's Travels in the Interior of North America, 1832–1834" (Cleveland: A. H. Clark Co., 1906), 393.
32. Bank of the Manhattan Company, "Ships and Shipping of Old New York," 43; Solomon, *Water*, 289.
33. There are only thirty-five locks on the Erie Canal today. www.eriecanal.org/locks.html. Ronald E. Shaw, *Canals for a Nation: The Canal Era in the United States, 1790–1860* (Lexington: University of Kentucky Press, 1990), 44, 47; Sheriff, *The Artificial River*, 67, 72, 79.

34. Chambers, *The Conquest of Cholera*, 63, 91; Shaw, *Canals for a Nation*, 47; John W. Percy, "Erie Canal: From Lockport to Buffalo," *Buffalo Architecture and History* (Buffalo: Western New York Heritage Institute of Canisius College, 1993).

35. Percy, "Erie Canal."

36. Solomon, *Water*, 228.

37. Chester G. Moore, "Globalization and the Law of Unintended Consequences: Rapid Spread of Disease Vectors via Commerce and Travel," Colorado State University, Fort Collins, ISAC meeting, June 2011; EPA, "Growth of International Trade and Transportation," www.epa.gov/oia/trade/transport.html; David Ozonoff and Lewis Pepper, "Ticket to Ride: Spreading Germs a Mile High," *The Lancet* 365, no. 9463 (2005): 917.

38. "Country Comparison: Airports," CIA, *The World Factbook*, 2013.

39. "Top 10 Biggest Ports in the World in 2011," *Marine Insight*, Aug. 11, 2011.

40. "Multi-modal Mainland Connections," 2013, www.hongkongairport.com.

41. Chris Taylor, "The Chinese Plague," *The Age*, May 4, 2003; Mike Davis, *The Monster at Our Door: The Global Threat of Avian Flu* (New York: Henry Holt, 2005), 70.

42. Nathan Wolfe, *The Viral Storm: The Dawn of a New Pandemic Age* (New York: Times Books, 2011), 160.

43. Christopher R. Braden et al., "Progress in Global Surveillance and Response Capacity 10 Years After Severe Acute Respiratory Syndrome," *Emerging Infectious Diseases* 19, no. 6 (2013): 864.

44. "What You Should Know About SARS," *The Vancouver Province*, March 23, 2003; Wolfe, *The Viral Storm*, 160; Forum on Microbial Threats, *Learning from SARS: Preparing for the Next Disease Outbreak* (Washington, DC: National Academies Press, 2004); Davis, *The Monster at Our Door*, 72–73.

45. Grady, "Ebola Cases Could Reach 1.4 Million"; David Kroll, "Nigeria Free of Ebola as Final Surveillance Contacts Are Released," *Forbes*, Sept. 23, 2014.

46. "India's Wealth Triples in a Decade to $3.5 Trillion," *The Economic Times* (India), Oct. 9, 2010.

47. "Medical Tourism in the Superbug Age," *The Times of India*, April 17, 2011.

48. "Medanta the Medicity," www.medanta.org/about_gallery.aspx.

49. Amit Sengupta and Samiran Nundy, "The Private Health Sector in India," *BMJ* 331, no. 7526 (Nov. 19, 2005): 1157–58; George K. Varghese et al., "Bacterial Organisms and Antimicrobial Resistance Patterns," *The Journal of the Association of Physicians of India* 58 supp. (December 2010): 23–24; Dawn Sievert et al., "Antimicrobial-Resistant Pathogens Associated with Healthcare-Associated Infections: Summary of Data Reported to the National Healthcare Safety Network at the Centers for Disease Control and Prevention, 2009–2010," *Infection Control and Hospital Epidemiology* 34, no. 1 (Jan. 2013): 1–14.

50. Maryn McKenna, "The Enemy Within," *Scientific American*, April 2011, 46–53; Chand Wattal et al., "Surveillance of Multidrug Resistant Organisms in Tertiary Care Hospital in Delhi, India," *The Journal of the Association of Physicians of India* 58 supp. (Dec. 2010): 32–36; Timothy R. Walsh and Mark A. Toleman, "The New Medical Challenge: Why NDM-1? Why Indian?" *Expert Review of Anti-Infective Therapy* 9, no. 2 (Feb. 2011): 137–41.

51. CDC, "Detection of Enterobacteriaceae Isolates Carrying Metallo-Beta-Lactamase—United States, 2010," June 25, 2010, www.cdc.gov/mmwr/preview

/mmwrhtml/mm5924a5.htm; Deverick J. Anderson, "Surgical Site Infections," *Infectious Disease Clinics of North America* 25, no. 1 (2011): 135–53; M. Berrazeg et al., "New Delhi Metallo-beta-lactamase Around the World: An eReview Using Google Maps," *Eurosurveillance* 19, no. 20 (2014).
52. Interview with Chand Wattal, Jan. 9, 2012.

3. FILTH
  1. Richard G. Feachem et al., *Sanitation and Disease: Health Aspects of Excreta and Wastewater Management*, World Bank Studies in Water Supply and Sanitation 3 (New York: John Wiley, 1983); Uno Winblad, "Towards an Ecological Approach to Sanitation," Swedish International Development Cooperation Agency, 1997.
  2. Rose George, *The Big Necessity: The Unmentionable World of Human Waste and Why It Matters* (New York: Metropolitan Books, 2008), 2.
  3. Joan H. Geismar, "Where Is Night Soil? Thoughts on an Urban Privy," *Historical Archaeology* 27, no. 2 (1993): 57–70; Laura Noren, *Toilet: Public Restrooms and the Politics of Sharing* (New York: NYU Press, 2010); Ewald, *Evolution of Infectious Disease*, 80.
  4. Katherine Ashenburg, *The Dirt on Clean: An Unsanitized History* (New York: North Point Press, 2007), 43; Solomon, *Water*, 251–53.
  5. George, *The Big Necessity*, 2.
  6. Ashenburg, *The Dirt on Clean*; Solomon, *Water*.
  7. Ashenburg, *The Dirt on Clean*, 94.
  8. Solomon, *Water*, 253.
  9. Ashenburg, *The Dirt on Clean*, 95, 100, 107.
 10. Martin V. Melosi, *The Sanitary City: Environmental Services in Urban America from Colonial Times to Present*, abridged ed. (Pittsburgh: University of Pittsburgh Press, 2008), 12.
 11. Benedetta Allegranzi et al., "Religion and Culture: Potential Undercurrents Influencing Hand Hygiene Promotion in Health Care," *American Journal of Infection Control* 37, no. 1 (2009): 28–34; Ashenburg, *The Dirt on Clean*, 59, 75.
 12. Echenberg, *Africa in the Time of Cholera*, 8.
 13. George, *The Big Necessity*, 8.
 14. John Duffy, *A History of Public Health in New York City 1625–1866* (New York: Russell Sage Foundation, 1968), 18; Gerard T. Koeppel, *Water for Gotham: A History* (Princeton: Princeton University Press, 2000), 12, 21.
 15. Melosi, *The Sanitary City*, 115.
 16. Tyler Anbinder, *Five Points: The 19th-Century New York City Neighborhood That Invented Tap Dance, Stole Elections, and Became the World's Most Notorious Slum* (New York: Plume, 2001), 74, 86.
 17. Eric W. Sanderson, *Manahatta: A Natural History of New York City* (New York: Harry N. Abrams, 2009), 215; Duffy, *A History of Public Health*, 185, 363.
 18. Duffy, *A History of Public Health*, 364.
 19. Asa Greene, *A Glance at New York: Embracing the City Government, Theatres, Hotels, Churches, Mobs, Monopolies, Learned Professions, Newspapers, Rogues, Dandies, Fires and Firemen, Water and Other Liquids, &c., &c.* (New York: A. Greene, 1837).

20. Argonne National Laboratory, "Cleaning Water Through Soil," Nov. 6, 2004, www.newton.dep.anl.gov/askasci/gen01/gen01688.htm.

21. Koeppel, *Water for Gotham*, 9; Sanderson, *Manahatta*; 87.

22. Greene, *A Glance at New York*.

23. Koeppel, *Water for Gotham*, 16, 52, 117.

24. Interview with Robert D. Mutch, Nov. 27, 2012; Duffy, *A History of Public Health*, 211; Nelson Manfred Blake, *Water for the Cities: A History of the Urban Water Supply Problem in the United States* (Syracuse, NY: Syracuse University Press, 1956), 124; "Old Water Tank Building Gives Way to Trade," *The New York Times*, July 12, 1914.

25. Blake, *Water for the Cities*, 126.

26. Koeppel, *Water for Gotham*, 64.

27. The data collected was in grains per gallon: 1 grain=64.8 mg; 1 gallon=3780 g. Koeppel, *Water for Gotham*, 121, 141.

28. J. S. Guthrie et al., "Alcohol and Its Influence on the Survival of *Vibrio cholerae*," *British Journal of Biomedical Science* 64, no. 2 (2007): 91–92.

29. Peter C. Baldwin, *In the Watches of the Night: Life in the Nocturnal City, 1820–1830* (Chicago: University of Chicago Press, 2012); Geismar, "Where Is Night Soil?"; Charles E. Rosenberg, *The Cholera Years: The United States in 1832, 1849, and 1866* (Chicago: University of Chicago Press, 1987), 112.

30. Documents of the Board of Aldermen of the City of New-York, vol. 9, document 18.

31. Sanderson, *Manahatta*, 10, 64, 153; Duffy, *A History of Public Health*, 25, 91, 379, 407; Feachem, *Sanitation and Disease*; Anbinder, *Five Points*, 87.

32. Duffy, *A History of Public Health*, 197.

33. Sanderson, *Manahatta*, 81.

34. Dudley Atkins, ed., *Reports of Hospital Physicians and Other Documents in Relation to the Epidemic Cholera of 1832* (New York: G. & C. & H. Carvill, 1832); James R. Manley, "Letters addressed to the Board of Health, and to Richard Riker, recorder of the city of New-York: on the subject of his agency in constituting a special medical council," Board of Health publication (New York: Peter van Pelt, 1832).

35. Steven Johnson, *The Ghost Map: The Story of London's Most Terrifying Epidemic— and How It Changed Science, Cities and the Modern World* (New York: Riverhead Books, 2006), 37.

36. Greene, *A Glance at New York*.

37. A drink consisting of water with just 15 percent gin would have to sit for twenty-six hours before cholera vibrio in it would perish. J. S. Guthrie et al., "Alcohol and Its Influence on the Survival of *Vibrio cholerae*," *British Journal of Biomedical Science* 64, no. 2 (2007): 91–92.

38. Mark Kurlansky, *The Big Oyster: History on the Half Shell* (New York: Random House, 2007); Duffy, *A History of Public Health*, 226.

39. Blake, *Water for the Cities*, 60.

40. "Extract of a letter from New-York, dated July 19, 1832," *The Liberator*, July 28, 1832; Atkins, *Reports of Hospital Physicians*.

41. Atkins, *Reports of Hospital Physicians*.

42. *The Cholera Bulletin, Conducted by an Association of Physicians*, vol. 1, nos. 1–24, 1832 (New York: Arno Press, 1972), 6.

43. Philip Hone, *The Diary of Philip Hone, 1828–1851* (New York: Dodd, Mead, 1910); John N. Ingham, *Biographical Dictionary of American Business Leaders*, vol. 1 (Santa Barbara, CA: Greenwood Publishing, 1983); Atkins, *Reports of Hospital Physicians.*

44. Atkins, *Reports of Hospital Physicians.*

45. Letter from Cornelia Laura Adams Tomlinson to Maria Annis Dayton and Cornelia Laura Tomlinson Weed, June 22, 1832, in "Genealogical Story (Dayton and Tomlinson)," told by Laura Dayton Fessenden (Cooperstown, NY: Crist, Scott & Parshall, 1902).

46. *Autobiography of N. T. Hubbard: With Personal Reminiscences of New York City from 1798 to 1875* (New York: J. F. Trow & Son, 1875).

47. Rosenberg, *The Cholera Years,* 32.

48. Hone, *The Diary of Philip Hone.*

49. Chris Swann, *A Survey of Residential Nutrient Behaviors in the Chesapeake Bay* (Ellicott City, MD: Chesapeake Research Consortium, Center for Watershed Protection, 1999).

50. Traci Watson, "Dog Waste Poses Threat to Water," *USA Today,* June 6, 2002.

51. Robert M. Bowers et al., "Sources of Bacteria in Outdoor Air Across Cities in the Midwestern United States," *Applied and Environmental Microbiology* 77, no. 18 (2011): 6350–56.

52. Dana M. Woodhall, Mark L. Eberhard, and Monica E. Parise, "Neglected Parasitic Infections in the United States: Toxocariasis," *The American Journal of Tropical Medicine and Hygiene* 90, no. 5 (2014): 810–13.

53. P. S. Craig et al., "An Epidemiological and Ecological Study of Human Alveolar Echinococcosis Transmission in South Gansu, China," *Acta Tropica* 77, no. 2 (2000): 167–77.

54. Jillian P. Fry et al., "Investigating the Role of State and Local Health Departments in Addressing Public Health Concerns Related to Industrial Food Animal Production Sites," *PLoS ONE* 8, no. 1 (2013): e54720.

55. JoAnn Burkholder et al., "Impacts of Waste from Concentrated Animal Feeding Operations on Water Quality," *Environmental Health Perspectives* 115, no. 2 (2007): 308.

56. Robbin Marks, "Cesspools of Shame: How Factory Farm Lagoons and Sprayfields Threaten Environmental and Public Health," Natural Resources Defense Council and the Clean Water Network, July 2001; Burkholder, "Impacts of Waste from Concentrated Animal Feeding Operations"; Wendee Nicole, "CAFOs and Environmental Justice: The Case of North Carolina," *Environmental Health Perspectives* 121, no. 6 (2013): a182–89.

57. Lee Bergquist and Kevin Crowe, "Manure Spills in 2013 the Highest in Seven Years Statewide," *Milwaukee Wisconsin Journal Sentinel,* Dec. 5, 2013; Peter T. Kilborn, "Hurricane Reveals Flaws in Farm Law," *The New York Times,* Oct. 17, 1999.

58. Xiuping Jiang, Jennie Morgan, and Michael P. Doyle, "Fate of *Escherichia coli* O157:H7 in Manure-Amended Soil," *Applied and Environmental Microbiology* 68, no. 5 (2002): 2605–609; Margo Chase-Topping et al., "Super-Shedding and the Link Between Human Infection and Livestock Carriage of *Escherichia coli* O157," *Nature Reviews Microbiology* 6, no. 12 (2008): 904–12; CDC, "*Escherichia coli* O157:H7, General Information—NCZVED," Jan. 6, 2011; J. A. Cotruvo et al., "Waterborne Zoonoses: Identification, Causes, and Control," WHO, 2004, 140.

59. NDM-1's capacity to move into bacterial species peaks at ambient, as opposed to body, temperatures. That may explain why it's already been found in environmental strains of both *Vibrio cholerae* and *Shigella boydii*, a culprit in severe dysentery. T. R. Walsh et al., "Dissemination of NDM-1 Positive Bacteria in the New Delhi Environment and Its Implications for Human Health: An Environmental Point Prevalence Study," *The Lancet Infectious Diseases* 11, no. 5 (2011): 355–62.

60. Drexler, *Secret Agents*, 146; McKenna, *Superbug*, 60–63; S. Tsubakishita et al., "Origin and Molecular Evolution of the Determinant of Methicillin Resistance in Staphylococci," *Antimicrobial Agents and Chemotherapy* 54, no. 10 (2010): 4352–59.

61. Maryn McKenna, "E. Coli: Some Answers, Many Questions Still," Wired.com, June 22, 2011; Yonatan H. Grad et al., "Comparative Genomics of Recent Shiga Toxin–Producing *Escherichia coli* O104:H4: Short-Term Evolution of an Emerging Pathogen," *mBio* 4, no. 1 (2013): e00452–12.

62. Ross Anderson, "Sprouts and Bacteria: It's the Growing Conditions," *Food Safety News*, June 6, 2011.

63. G. Gault et al., "Outbreak of Haemolytic Uraemic Syndrome and Bloody Diarrhoea Due to *Escherichia coli* O104:H4, South-West France, June 2011," *Eurosurveillance* 16, no. 26 (2011).

64. McKenna, "E. Coli: Some Answers; "'A Totally New Disease Pattern': Doctors Shaken by Outbreak's Neurological Devastation," Spiegel Online, June 9, 2011; Gault, "Outbreak of Haemolytic Uraemic Syndrome."

65. Ralf P. Vonberg et al., "Duration of Fecal Shedding of Shiga Toxin–Producing *Escherichia coli* O104:H4 in Patients Infected During the 2011 Outbreak in Germany: A Multicenter Study," *Clinical Infectious Diseases* 56 (2013).

66. Haiti Grassroots Watch, "Behind the Cholera Epidemic—Excreta," December 21, 2010.

67. George, *The Big Necessity*, 89, 99.

68. Solomon, *Water*, 265.

69. Interview with Brian Concannon, July 23, 2013.

70. Haiti Grassroots Watch, "Behind the Cholera Epidemic."

71. Associated Press interview, "UN Envoy Farmer Says Haiti Cholera Outbreak Is Now World's Worst," Oct. 18, 2011.

72. Walsh, "Dissemination of NDM-1 Positive Bacteria."

73. In January 2011, a man in Hong Kong was found to be infected with an *E. coli* strain with NDM-1. With no history of hospitalizations, experts suspect he may have picked up the infection from excreta-contaminated waters or soils. In May 2011, NDM-1 was detected in a patient in Canada. This eighty-six-year-old man hadn't left western Ontario in at least ten years. A polluted local environment may have exposed him to the bug, too. McKenna, "The Enemy Within"; J. V. Kus et al., "New Delhi Metallo-ss-lactamase-1: Local Acquisition in Ontario, Canada, and Challenges in Detection," *Canadian Medical Association Journal* 183, no. 11 (Aug. 9, 2011): 1257–61.

4. CROWDS

1. Despite not being noticeably ill, carriers could still unknowingly contribute to cholera's spread, excreting as much as 500,000,000 cholera vibrio every day. (Calculated as 1,000,000 cholera vibrio per gram of feces, with each person producing

500 grams of feces/day.) Feachem, *Sanitation and Disease*. C. T. Codeço, "Endemic and Epidemic Dynamics of Cholera: The Role of the Aquatic Reservoir," *BMC Infectious Diseases* 1, no. 1 (2001); Atkins, *Reports of Hospital Physicians*.

2. Cholera immunity is understood to be long-lasting but the mechanisms behind it remain unclear. Eric J. Nelson et al., "Cholera Transmission: The Host, Pathogen and Bacteriophage Dynamic," *Nature Reviews Microbiology* 7, no. 10 (2009): 693–702.

3. Rosenberg, *The Cholera Years*, 35.

4. James D. Oliver, "The Viable but Nonculturable State in Bacteria," *The Journal of Microbiology* 43, no. 1 (2005): 93–100.

5. Anbinder, *Five Points*, 14–27; Ashenburg, *The Dirt on Clean*, 178; Richard Plunz, *A History of Housing in New York City* (New York: Columbia University Press, 1990).

6. Simon Szreter, "Economic Growth, Disruption, Deprivation, Disease, and Death: On the Importance of the Politics of Public Health for Development," *Population and Development Review* 23 (1997): 693–728.

7. John Reader, *Potato: A History of the Propitious Esculent* (New Haven: Yale University Press, 2009).

8. Ian Steadman, "Mystery Irish Potato Famine Pathogen Identified 170 Years Later," *Wired UK*, May 21, 2013.

9. Reader, *Potato*; Everett M. Rogers, *Diffusion of Innovations*, 5th ed. (New York: Free Press, 2003), 452; W. C. Paddock, "Our Last Chance to Win the War on Hunger," *Advances in Plant Pathology* 8 (1992), 197–222.

10. Duffy, *A History of Public Health*, 273.

11. Cormac Ó. Gráda and Kevin H. O'Rourke, "Migration as Disaster Relief: Lessons from the Great Irish Famine," *European Review of Economic History* 1, no. 1 (1997): 3–25.

12. Jacob A. Riis, *How the Other Half Lives: Studies Among the Tenements of New York*, ed. David Leviatin (New York: St. Martin's Press, 1996 [1890]), 67; Anbinder, *Five Points*, 74.

13. Anbinder, *Five Points*, 81.

14. Plunz, *A History of Housing in New York City*.

15. Anbinder, *Five Points*, 74–77.

16. Riis, *How the Other Half Lives*, 65.

17. Anbinder, *Five Points*, 14–27, 69, 71, 74–79, 175, 306; Rosenberg, *The Cholera Years*, 34.

18. Davis, *The Monster at Our Door*, 154.

19. Koeppel, *Water for Gotham*, 287.

20. Rosenberg, *The Cholera Years*, 104, 106, 113–14, 121, 145; Anbinder, *Five Points*, 119.

21. Michael R. Haines, "The Urban Mortality Transition in the United States, 1800–1940," National Bureau of Economic Research Historical Paper no. 134, July 2001; Michael Haines, "Health, Height, Nutrition and Mortality: Evidence on the 'Antebellum Puzzle' from Union Army Recruits for New York State and the United States," in John Komlos and Jörg Baten, eds., *The Biological Standard of Living in Comparative Perspective* (Stuttgart: Franz Steiner Verlag, 1998); Robert Woods, "Urban-Rural Mortality Differentials: An Unresolved Debate," *Population and Development Review* 29, no. 1 (2003): 29–46.

22. Woods, "Urban-Rural Mortality Differentials."
23. Duffy, A History of Public Health, 291.
24. Adam Gopnik, "When Buildings Go Up, the City's Distant Past Has a Way of Resurfacing," The New Yorker, Feb. 4, 2002; Michael O. Allen, "5 Points Had Good Points," Daily News, Feb. 22, 1998.
25. G. T. Kingsley, "Housing, Health, and the Neighborhood Context," American Journal of Preventive Medicine 4, supp. 3 (April 2003): 6–7.
26. Davis, The Monster at Our Door, 154.
27. Nature Conservancy, "Global Impact of Urbanization Threatening World's Biodiversity and Natural Resources," ScienceDaily, June 2008.
28. Davis, The Monster at Our Door, 152.
29. Danielle Nierenberg, "Factory Farming in the Developing World," World Watch magazine 16, no. 3 (May/June 2003).
30. Xavier Pourrut et al., "The Natural History of Ebola Virus in Africa," Microbes and Infection 7, no. 7 (2005): 1005–14.
31. E. M., Leroy, J. P. Gonzalez, and S. Baize, "Ebola and Marburg Haemorrhagic Fever Viruses: Major Scientific Advances, but a Relatively Minor Public Health Threat for Africa," Clinical Microbiology and Infection 17, no. 7 (2011): 964–76.
32. Todd C. Frankel, "It Was Already the Worst Ebola Outbreak in History. Now It's Moving into Africa's Cities," The Washington Post, Aug. 30, 2014; "Ebola Virus Reaches Guinea's Capital Conakry," Al Jazeera, March 28, 2014; "Seven Die in Monrovia Ebola Outbreak," BBC News, June 17, 2014; "Sierra Leone Capital Now in Grip of Ebola," Al Jazeera, Aug. 6, 2014.
33. The transmission rate was not found to have increased in the capital city of Sierra Leone, however, for reasons that are unclear. S. Towers, O. Patterson-Lomba, and Chavez C. Castillo, "Temporal Variations in the Effective Reproduction Number of the 2014 West Africa Ebola Outbreak," PLoS Currents Outbreaks, Sept. 18, 2014.
34. Interview with James Lloyd-Smith, Nov. 30, 2011.
35. Frankel, "It Was Already the Worst Ebola Outbreak."
36. Barry S. Hewlett and Bonnie L. Hewlett, Ebola, Culture and Politics: The Anthropology of an Emerging Disease (Belmont, CA: Thomson Wadsworth, 2008), 55.
37. Paul W. Ewald, Plague Time: How Stealth Infections Cause Cancers, Heart Disease, and Other Deadly Ailments (New York: Simon and Schuster, 2000), 25.
38. "Pathogen Safety Data Sheet: Infectious Substances: Mycobacterium Tuberculosis Complex," Public Health Agency of Canada, Oct. 6, 2014; Michael Z. David and Robert S. Daum, "Community-Associated Methicillin-Resistant Staphylococcus aureus: Epidemiology and Clinical Consequences of an Emerging Epidemic," Clinical Microbiology Reviews 23, no. 3 (2010): 616–87.
39. Lise Wilkinson and A. P. Waterson, "The Development of the Virus Concept as Reflected in Corpora of Studies on Individual Pathogens: 2. The Agent of Fowl Plague—A Model Virus?" Medical History 19 (1975): 52–72; Sander Herfst et al., "Airborne Transmission of Influenza A/H5N1 Virus Between Ferrets," Science 336, no. 6088 (2012): 1534–41; Dennis J. Alexander, "An Overview of the Epidemiology of Avian Influenza," Vaccine 25, no. 30 (2007): 5637–44.
40. Yohei Watanabe, Madiha S. Ibrahim, and Kazuyoshi Ikuta, "Evolution and Control of H5N1," EMBO Reports 14, no. 2 (2013): 117–22.
41. Les Sims and Clare Narrod, Understanding Avian Influenza: A Review of the Emer-

gence, Spread, Control, Prevention and Effects of Asian-Lineage H5N1 Highly Pathogenic Viruses (Rome: FAO, 2007).

42. James Truscott et al., "Control of a Highly Pathogenic H5N1 Avian Influenza Outbreak in the GB Poultry Flock," *Proceedings of the Royal Society B* 274 (2007): 2287–95.

43. M. S. Beato and I. Capua, "Transboundary Spread of Highly Pathogenic Avian Influenza Through Poultry Commodities and Wild Birds: A Review," *Revue Scientifique et Technique (International Office of Epizootics)* 30, no. 1 (April 2011): 51–61.

44. Shefali Sharma et al., eds., *Fair or Fowl? Industrialization of Poultry Production in China, Global Meat Complex* (Minneapolis: Institute for Agriculture and Trade Policy, February 2014).

45. S. P. Cobb, "The Spread of Pathogens Through Trade in Poultry Meat: Overview and Recent Developments," *Revue Scientifique et Technique (International Office of Epizootics)* 30, no. 1 (April 2011): 149–64.

46. Truscott, "Control of a Highly Pathogenic H5N1 Avian Influenza Outbreak."

47. Alexander, "An Overview of the Epidemiology of Avian Influenza."

48. Cobb, "The Spread of Pathogens Through Trade in Poultry Meat."

49. Beato and Capua, "Transboundary Spread of Highly Pathogenic Avian Influenza"; interview with Malik Peiris, Jan. 17, 2012.

50. Debby Van Riel et al., "H5N1 Virus Attachment to Lower Respiratory Tract," *Science* 312, no. 5772 (2006): 399.

51. Interview with Malik Peiris.

52. Beato and Capua. "Transboundary Spread of Highly Pathogenic Avian Influenza"; interview with Malik Peiris.

53. A. Marm Kilpatrick et al., "Predicting the Global Spread of H5N1 Avian Influenza," *Proceedings of the National Academy of Sciences* 103, no. 51 (2006): 19368–73.

54. Scientists suspect that's because H5N1 can't yet bind well to the easily accessible cells in our upper respiratory tracts. (It does bind to the cells in our lower respiratory tracts, including our lungs, which is why it makes us so sick.) Watanabe, Ibrahim, and Ikuta, "Evolution and Control of H5N1"; World Health Organization, "Cumulative Number of Confirmed Human Cases for Avian Influenza A(H5N1) Reported to WHO, 2003–2014," July 27, 2014.

55. Sims and Narrod, *Understanding Avian Influenza.*

56. Watanabe, Ibrahim, and Ikuta, "Evolution and Control of H5N1."

57. Kevin Drew, "China Says Man Dies from Bird Flu," *The New York Times*, Dec. 31, 2011.

58. Davis, *The Monster at Our Door*, 181.

59. Donald G. McNeil, "A Flu Epidemic That Threatens Birds, Not Humans," *The New York Times*, May 4, 2015.

60. Wenjun Ma, Robert E. Kahn, and Juergen A. Richt, "The Pig as a Mixing Vessel for Influenza Viruses: Human and Veterinary Implications," *Journal of Molecular and Genetic Medicine* 3, no. 1 (2009): 158.

61. Davis, *The Monster at Our Door*, 17.

62. Mindi Schneider, "Feeding China's Pigs: Implications for the Environment, Chi-

na's Smallholder Farmers and Food Security," Institute for Agriculture and Trade Policy, May 2011.

63. S. McOrist, K. Khampee, and A. Guo, "Modern Pig Farming in the People's Republic of China: Growth and Veterinary Challenges," *Revue Scientifique et Technique (International Office of Epizootics)* 30, no. 3 (2011): 961–68.

64. Qiyun Zhu et al., "A Naturally Occurring Deletion in Its NS Gene Contributes to the Attenuation of an H5N1 Swine Influenza Virus in Chickens," *Journal of Virology* 82, no. 1 (2008): 220–28.

65. Michael Osterholm, "This Year, It Seems, It's 'Risk On' with Swine Flu," *StarTribune* (Minneapolis), Aug. 26, 2012.

66. Department of Health and Human Services, "H3N2v," flu.gov/about_the_flu/h3n2v.

67. Maura Lerner and Curt Brown, "Will New Flu Strain Close the Swine Barn at Minnesota State Fair?" *StarTribune*, Aug. 21, 2012.

68. Di Liu et al., "Origin and Diversity of Novel Avian Influenza A H7N9 Viruses Causing Human Infection: Phylogenetic, Structural, and Coalescent Analyses," *The Lancet* 381, no. 9881 (2013): 1926–32; Rongbao Gao et al., "Human Infection with a Novel Avian-Origin Influenza A (H7N9) Virus," *The New England Journal of Medicine* 368, no. 20 (2013): 1888–97; Yu Chen et al., "Human Infections with the Emerging Avian Influenza A H7N9 Virus from Wet Market Poultry: Clinical Analysis and Characterisation of Viral Genome," *The Lancet* 381, no. 9881 (2013): 1916–25; Hongjie Yu et al., "Effect of Closure of Live Poultry Markets on Poultry-to-Person Transmission of Avian Influenza A H7N9 Virus: An Ecological Study," *The Lancet* 383, no. 9916 (2014): 541–48; Tokiko Watanabe et al., "Pandemic Potential of Avian Influenza A (H7N9) Viruses," *Trends in Microbiology* 22, no. 11 (2014): 623–31.

5. CORRUPTION

1. Hewlett and Hewlett, *Ebola, Culture, and Politics*, 44–45.

2. Ernst Fehr, Urs Fischbacher, and Simon Gächter, "Strong Reciprocity, Human Cooperation, and the Enforcement of Social Norms," *Human Nature* 13, no. 1 (2002): 1–25; Eric Michael Johnson, "Punishing Cheaters Promotes the Evolution of Cooperation," *The Primate Diaries* (*Scientific American* blog), Aug. 16, 2012.

3. Koeppel, *Water for Gotham*, 80; Beatrice G. Reubens, "Burr, Hamilton and the Manhattan Company: Part I: Gaining the Charter," *Political Science Quarterly* 72, no. 4 (1957): 578–607; Solomon, *Water*, 254–55; Fairmount Water Works Interpretive Center, fairmountwaterworks.org.

4. Blake, *Water for the Cities*, 48, 143.

5. David O. Stewart, "The Perils of Nonpartisanship: The Case of Aaron Burr," *The Huffington Post*, Sept. 14, 2011.

6. Koeppel, *Water for Gotham*, 36.

7. Reubens, "Burr, Hamilton and the Manhattan Company: Part I."

8. Koeppel, *Water for Gotham*, 82–83.

9. Blake, *Water for the Cities*, 73.

10. Koeppel, *Water for Gotham*, 87.

11. Beatrice G. Reubens, "Burr, Hamilton and the Manhattan Company: Part II: Launching a Bank," *Political Science Quarterly* 73, no. 1 (1958): 100–125.

12. Blake, *Water for the Cities*, 60.

13. Reubens, "Burr, Hamilton and the Manhattan Company: Part II."

14. Koeppel, *Water for Gotham*, 87.

15. Blake, *Water for the Cities*, 106.

16. Reubens, "Burr, Hamilton and the Manhattan Company: Part II."

17. Subhabrata Bobby Banerjee, "Corporate Social Responsibility: The Good, the Bad and the Ugly," *Critical Sociology* 34, no. 1 (2008): 51–79.

18. Blake, *Water for the Cities*, 102.

19. Purchasing power of $9,000 in 1800 is equal to $167,445 in current dollars, according to "Historical Currency Conversions," http://futureboy.us/fsp/dollar.fsp?quantity =9000&currency=dollars&fromYear=1800; Koeppel, *Water for Gotham*, 100.

20. Reubens, "Burr, Hamilton and the Manhattan Company: Part I."

21. Koeppel, *Water for Gotham*, 99.

22. Reubens, "Burr, Hamilton and the Manhattan Company: Part I."

23. "The History of JPMorgan Chase & Co.," www.jpmorganchase.com/corporate /About-JPMC/jpmorgan-history.

24. Blake, *Water for the Cities*, 68.

25. Melosi, *The Sanitary City*, 16.

26. Blake, *Water for the Cities*, 77.

27. Howard Markel, *When Germs Travel: Six Major Epidemics That Have Invaded America Since 1900 and the Fears They Have Unleashed* (New York: Pantheon, 2004), 51.

28. Frank M. Snowden, *Naples in the Time of Cholera, 1884–1911* (New York: Cambridge University Press, 1995), 80.

29. Ibid., 80–81.

30. Delaporte, *Disease and Civilization*, 194.

31. Duffy, *A History of Public Health*, 119.

32. Ibid., 134.

33. Chambers, *The Conquest of Cholera*, 105.

34. Erwin H. Ackerknecht, "Anticontagionism Between 1821 and 1867," *International Journal of Epidemiology* 38, no. 1 (2009): 7–21.

35. Delaporte, *Disease and Civilization*, 140.

36. Ackerknecht, "Anticontagionism Between 1821 and 1867."

37. Manley, "Letters addressed to the Board of Health."

38. Confusingly, these two schools of thought about disease causation were defined as "infection" on one hand and "contagion" on the other. An infection, from the Latin *inficere*, "to stain," was a disease that traveled in stinky airs, staining the body with sickness the way newly developed and highly odoriferous chemical dyes stained fabric. The older concept of "contagion" referred to diseases that spread from person to person, like a seed passed from plant to plant. The term derives from the Latin for "contact with filth." Delaporte, *Disease and Civilization*, 182; Snowden, *Naples in the Time of Cholera*, 68.

39. Rosenberg, *The Cholera Years*, 41.

40. Duffy, *A History of Public Health*, 161, 330–31.

41. Rosenberg, *The Cholera Years*, 104; Echenberg, *Africa in the Time of Cholera*, 76; Duffy, *A History of Public Health*, 166.

42. Tuite, Chan, and Fisman, "Cholera, Canals, and Contagion."

43. Ibid.

44. Transactions of the Medical Society of the State of New York, vol. 1 (Albany, 1833).

45. Rosenberg, The Cholera Years, 98; Delaporte, Disease and Civilization, 111.

46. Percy, "Erie Canal."

47. Chambers, The Conquest of Cholera, 39.

48. Rosenberg, The Cholera Years, 20, 26.

49. The Cholera Bulletin, vol. 1, nos. 2 and 3, 1832.

50. Rosenberg, The Cholera Years, 25.

51. Snowden, Naples in the Time of Cholera, 197–98, 301–309, 316–57.

52. Davis, The Monster at Our Door, 69–70.

53. Richard Wenzel, "International Perspectives on Infection Control in Healthcare Institutions," International Conference on Emerging Infectious Diseases, Atlanta, GA, March 12, 2012.

54. Davis, The Monster at Our Door, 69–75.

55. Juan O. Tamayo, "Cuba Stays Silent About Deadly Cholera Outbreak," The Miami Herald, Dec. 8, 2012.

56. George, The Big Necessity, 213.

57. Jennifer Yang, "How Medical Sleuths Stopped a Deadly New SARS-like Virus in Its Tracks," Toronto Star, Oct. 21, 2012.

58. Tom Clark, "Drug Resistant Superbug Threatens UK Hospitals," Channel 4 News, Oct. 28, 2010.

59. Interview with Timothy Walsh, Dec. 21, 2011.

60. www.globalpolicy.org/component/content/article/221/47211.html.

61. Patricia Cohen, "Oxfam Study Finds Richest 1% Is Likely to Control Half of Global Wealth by 2016," The New York Times, Jan. 19, 2015.

62. Alexander Fleming, "Penicillin," Nobel lecture, Dec. 11, 1945, www.nobelprize .org/nobel_prizes/medicine/laureates/1945/fleming-lecture.pdf.

63. Spellberg, "Antimicrobial Resistance."

64. Center for Veterinary Medicine, "Summary Report on Antimicrobials Sold or Distributed for Use in Food-Producing Animals," FDA, Sept. 2014.

65. Walsh and Toleman, "The New Medical Challenge."

66. Clark, "Drug Resistant Superbug Threatens UK Hospitals"; Global Antibiotic Resistance Partnership (GARP)-India Working Group, "Rationalizing Antibiotic Use to Limit Antibiotic Resistance in India," The Indian Journal of Medical Research (Sept. 2011): 281–94.

67. D. M. Livermore, "Has the Era of Untreatable Infections Arrived?" The Journal of Antimicrobial Chemotherapy 64, supp. 1 (2009): i29–i36; T. R. Walsh, "Emerging Carbapenemases: A Global Perspective," International Journal of Antimicrobial Agents 36 supp. 3 (2010): s8–s14.

68. Washer, Emerging Infectious Diseases; David and Daum, "Community-Associated Methicillin-Resistant Staphylococcus aureus"; McKenna, Superbug, 160.

69. Drexler, Secret Agents, 152–54.

70. Sara Reardon, "FDA Institutes Voluntary Rules on Farm Antibiotics," Nature News, Dec. 11, 2013.

71. McKenna, Superbug, 166.

72. Sara Reardon, "White House Takes Aim at Antibiotic Resistance," *Nature News*, Sept. 18, 2014.

73. Livermore, "Has the Era of Untreatable Infections Arrived?"

74. Michelle Bahrain et al., "Five Cases of Bacterial Endocarditis After Furunculosis and the Ongoing Saga of Community-Acquired Methicillin-Resistant *Staphylococcus aureus* Infections," *Scandinavian Journal of Infectious Diseases* 38, no. 8 (2006): 702–707.

75. G. R. Nimmo, "USA300 Abroad: Global Spread of a Virulent Strain of Community-Associated Methicillin-Resistant *Staphylococcus aureus*," *Clinical Microbiology and Infection* 18, no. 8 (2012): 725–34.

76. David and Daum, "Community-Associated Methicillin-Resistant *Staphylococcus aureus*."

77. Bahrain, "Five Cases of Bacterial Endocarditis."

78. Livermore, "Has the Era of Untreatable Infections Arrived?"

79. Pollack, "Looking for a Superbug Killer."

80. McKenna, "The Enemy Within."

81. Peter Utting et al., "UN-Business Partnerships: Whose Agenda Counts?" *Transnational Associations*, Dec. 8, 2000, 18.

82. J. Patrick Vaughan et al., "WHO and the Effects of Extrabudgetary Funds: Is the Organization Donor Driven?" *Health Policy and Planning* 11, no. 3 (1996); World Health Organization, "Programme Budget 2014–2015," www.who.int, May 24, 2013.

83. Sheri Fink, "WHO Leader Describes the Agency's Ebola Operations," *The New York Times*, Sept. 4, 2014.

84. Stuckler et al., "WHO's Budgetary Allocations and Burden of Disease: A Comparative Analysis," *The Lancet* 372 (2008): 9649.

85. Buse et al., "Public-Private Health Partnerships: A Strategy for WHO," *Bulletin of the World Health Organization* 79, no. 8 (2001): 748–54.

86. Maria Cheng and Raphael Satter, "Emails Show the World Health Organization Intentionally Delayed Calling Ebola a Public Health Emergency," Associated Press, March 20, 2015; Sarah Boseley, "World Health Organization Admits Botching Response to Ebola Outbreak," *The Guardian*, Oct. 17, 2014.

87. Andrew Bowman, "The Flip Side to Bill Gates' Charity Billions," *New Internationalist*, April 2012.

88. Sonia Shah, "Guerrilla War on Malaria," *Le Monde Diplomatique*, April 2011.

89. Some experts have raised questions about the Gates Foundation's investments in processed food and pharmaceutical companies. David Stuckler, Sanjay Basu, and Martin McKee, "Global Health Philanthropy and Institutional Relationships: How Should Conflicts of Interest be Addressed?" *PLoS Medicine* 8, no. 4 (2011): e1001020.

6. BLAME

1. Dan Coughlin, "WikiLeaks Haiti: US Cables Paint Portrait of Brutal, Ineffectual and Polluting UN Force," *The Nation*, Oct. 6, 2011.

2. Kathie Klarreich, "Will the United Nations' Legacy in Haiti Be All About Scandal?" *The Christian Science Monitor*, June 13, 2012.

3. "Fearful Crowds Wreck Clinic as Panic over Cholera Grows," *The Times* (London), Oct. 29, 2010.

4. "Oxfam Workers Flee Riot-Torn Cholera City as Disease Spreads Across Border," *The Times* (London), Nov. 17, 2010.

5. Samuel Cohn, "Pandemics: Waves of Disease, Waves of Hate from the Plague of Athens to AIDS," *Historical Research* 85, no. 230 (2012): 535–55.

6. Susan Sontag, *Illness as Metaphor and AIDS and Its Metaphors* (New York: Macmillan, 2001), 40–41.

7. Cohn, "Pandemics."

8. United Nations Senior Advisory Group, "Report of the Senior Advisory Group on Rates of Reimbursement to Troop-Contributing Countries and Other Related Issues," Oct. 11, 2012.

9. Zachary K. Rothschild et al., "A Dual-Motive Model of Scapegoating: Displacing Blame to Reduce Guilt or Increase Control," *Journal of Personality and Social Psychology* 102, no. 6 (2012): 1148.

10. Daniel Sullivan et al., "An Existential Function of Enemyship: Evidence That People Attribute Influence to Personal and Political Enemies to Compensate for Threats to Control," *Journal of Personality and Social Psychology* 98, no. 3 (2010): 434–49.

11. Rothschild, "A Dual-Motive Model of Scapegoating."

12. Neel L. Burton, *Hide and Seek: The Psychology of Self-Deception* (Oxford: Acheron Press, 2012).

13. Attila Pók, "Atonement and Sacrifice: Scapegoats in Modern Eastern and Central Europe," *East European Quarterly* 32, no. 4 (1998): 531.

14. Snowden, *Naples in the Time of Cholera*, 151.

15. Rosenberg, *The Cholera Years*, 33.

16. William J. Callahan, *Church, Politics, and Society in Spain, 1750–1874* (Cambridge, MA: Harvard University Press, 1984).

17. Rosenberg, *The Cholera Years*, 135.

18. Chambers, *The Conquest of Cholera*, 41.

19. Percy, "Erie Canal."

20. Rosenberg, *The Cholera Years*, 62–63.

21. William Watson, "The Sisters of Charity, the 1832 Cholera Epidemic in Philadelphia, and Duffy's Cut," *U.S. Catholic Historian* 27, no. 4 (Fall 2009): 1–16; Dan Barry, "With Shovels and Science, a Grim Story Is Told," *The New York Times*, March 24, 2013.

22. Barry, "With Shovels and Science."

23. W. Omar, "The Mecca Pilgrimage," *Postgraduate Medical Journal* 28, no. 319 (1952): 269.

24. M. C. Low, "Empire and the Hajj: Pilgrims, Plagues, and Pan-Islam Under British Surveillance, 1865–1908," *International Journal of Middle East Studies* 40, no. 2 (2008): 1–22.

25. F. E. Peters, *The Hajj: The Muslim Pilgrimage to Mecca and the Holy Places* (Princeton: Princeton University Press, 1994).

26. Valeska Huber, "The Unification of the Globe by Disease? The International Sanitary Conferences on Cholera, 1851–1894," *The Historical Journal* 49, no. 02 (2006): 453.

27. Low, "Empire and the Hajj."

28. Echenberg, *Africa in the Time of Cholera*, 37.

29. Harriet Moore, "Contagion from Abroad: U.S. Press Framing of Immigrants and

Epidemics, 1891 to 1893" (master's thesis, Georgia State University, Department of Communications, 2008), 1–113.

30. Howard Markel, *Quarantine! East European Jewish Immigrants and the New York City Epidemics of 1892* (Baltimore: Johns Hopkins University Press, 1997), 111–19.

31. Cohn, "Pandemics"; Rosenberg, *The Cholera Years*, 67.

32. "Death and Disbelievers," *The Economist*, Aug. 2, 2014; "Ebola: Guineans Riot in Nzerekore over Disinfectant," BBC News Africa, Aug. 29, 2014; Abby Phillip, "Eight Dead in Attack on Ebola Team in Guinea," *The Washington Post*, Sept. 28, 2014; Terrence McCoy, "Why the Brutal Murder of Several Ebola Workers May Hint at More Violence to Come," *The Washington Post*, Sept. 19, 2014.

33. Laurie Garrett, *The Coming Plague: Newly Emerging Diseases in a World out of Balance* (New York: Macmillan, 1994), 352.

34. Sonia Shah, *The Body Hunters: Testing New Drugs on the World's Poorest Patients* (New York: New Press, 2012), 104.

35. Pride Chigwedere et al., "Estimating the Lost Benefits of Antiretroviral Drug Use in South Africa," *JAIDS Journal of Acquired Immune Deficiency Syndromes* 49, no. 4 (2008): 410–15.

36. Gregory M. Herek and Eric K. Glunt, "An Epidemic of Stigma: Public Reactions to AIDS," *American Psychologist* 43, no. 11 (1988): 886.

37. Gregory M. Herek, "AIDS and Stigma," *American Behavioral Scientist* 42, no. 7 (1999): 1106–16; Mirko D. Grmek, *History of AIDS: Emergence and Origin of a Modern Pandemic* (Princeton: Princeton University Press, 1990); Paul Farmer, "Social Inequalities and Emerging Infectious Diseases," *Emerging Infectious Diseases* 2, no. 4 (1996): 259.

38. Edwidge Danticat, "Don't Let New AIDS Study Scapegoat Haitians," *The Progressive*, Nov. 7, 2007.

39. Washer, *Emerging Infectious Diseases*, 131–32.

40. Richard Preston, "West Nile Mystery," *The New Yorker*, Oct. 18, 1999.

41. Ibid.

42. "Chinese Refugees Face SARS Discrimination," CBC News, April 5, 2003; "China Syndrome," *The Economist*, April 10, 2003.

43. "Chinese Refugees Face SARS Discrimination"; "China Syndrome."

44. Chinese Canadian National Council—National Office, "Yellow Peril Revisited: Impact of SARS on the Chinese and Southeast Asian Communities," June 2004.

45. Robert Samuels Morello, "At Rock Creek Park, Harvesting Deer and Hard Feelings," *The Washington Post*, March 30, 2013.

46. "Are Deer the Culprit in Lyme Disease?" *The New York Times*, July 29, 2009.

47. Pam Belluck, "Tick-Borne Illnesses Have Nantucket Considering Some Deer-Based Solutions," *The New York Times*, Sept. 6, 2009.

48. Leslie Lake, "Former Norwalk Man Hunts Deer in New Reality Television Show," *The Hour*, April 21, 2013.

49. Ernesto Londo, "Egypt's Garbage Crisis Bedevils Morsi," *The Washington Post*, Aug. 27, 2012; "Swine Flu Pig Cull Destroys Way of Life for City's Coptic Rubbish Collectors," *The Times* (London), June 6, 2009; "For Egypt's Christians, Pig Cull Has Lasting Effects," *The Christian Science Monitor*, Sept. 3, 2009; "New Film Reveals the Story of Egyptian Trash Collectors," *Waste & Recycling News*, Jan. 23, 2012; "Copts Between the Rock of Islamism and a Hard Place," *The Times* (London),

Nov. 14, 2009; Michael Slackman, "Belatedly, Egypt Spots Flaws in Wiping Out Pigs," *The New York Times*, Sept. 19, 2009; "President Under Pressure to Solve Cairo's Trash Problems," *The New Zealand Herald*, Sept. 3, 2012.

50. Elisha P. Renne, *The Politics of Polio in Northern Nigeria* (Bloomington: Indiana University Press, 2010), 11, 40.

51. Declan Walsh, "Taliban Block Vaccinations in Pakistan," *The New York Times*, June 18, 2012.

52. Y. Paul and A. Dawson, "Some Ethical Issues Arising From Polio Eradication Programmes in India," *Bioethics* 19, no. 4 (2005): 393–406; Robert Fortner, "Polio in Retreat: New Cases Nearly Eliminated Where Virus Once Flourished," *Scientific American*, Oct. 28, 2010.

53. Declan Walsh, "Polio Crisis Deepens in Pakistan, With New Cases and Killings," *The New York Times*, Nov. 26, 2014.

54. Paul Greenough, "Intimidation, Coercion and Resistance in the Final Stages of the South Asian Smallpox Eradication Campaign, 1973–1975," *Social Science & Medicine* 41, no. 5 (1995): 633–45.

55. Michael Willrich, *Pox: An American History* (New York: Penguin Press, 2011), 118.

56. "How the CIA's Fake Vaccination Campaign Endangers Us All," *Scientific American*, May 3, 2013.

57. "Congo Republic Declares Polio Emergency," *The New York Times*, Nov. 9, 2010, 1–3.

58. WHO Global Alert and Response, "China: WHO Confirmation," Sept. 1, 2011, www.who.int/csr/don/2011_09_01/en/index.html; "WHO: Pakistan Polio Strain in Syria," Radio Free Europe, Nov. 12, 2013.

59. Donald G. McNeil, "Polio's Return After Near Eradication Prompts a Global Health Warning," *The New York Times*, May 5, 2014.

60. Saad B. Omer et al., "Vaccine Refusal, Mandatory Immunization, and the Risks of Vaccine-Preventable Diseases," *The New England Journal of Medicine* 360 (May 7, 2009): 1981–85; "Chinese CDC Admits Vaccine Reactions Cause Paralysis in Chinese Children," The Refusers, Oct. 10, 2013; Greg Poland, "Improving Adult Immunization and the Way of Sophia: A 12-Step Program," International Conference on Emerging Infectious Diseases, March 12, 2012, Atlanta, GA.

61. Warren Jones and Ami Klin, "Attention to Eyes Is Present but in Decline in 2–6-Month-Old Infants Later Diagnosed with Autism," *Nature*, Nov. 6, 2013.

62. Paul A. Offit, "Why Are Pharmaceutical Companies Gradually Abandoning Vaccines?" *Health Affairs*, May 2005.

63. "A Pox on My Child: Cool," *The Washington Post*, Sept. 20, 2005.

64. Omer, "Vaccine Refusal, Mandatory Immunization, and the Risks of Vaccine-Preventable Diseases."

65. Poland, "Improving Adult Immunization and the Way of Sophia."

66. Daniel Salmon et al., "Factors Associated with Refusal of Childhood Vaccines Among Parents of School-Aged Children," *JAMA Pediatrics* 159, no. 5 (May 2005): 470–76.

67. Mike Stobbe, "More Kids Skip School Shots in 8 States," Associated Press, Nov. 28, 2011.

68. CDC, "Notes from the Field: Measles Outbreak—Indiana, June–July 2011"; CDC, "U.S. Multi-State Measles Outbreak 2014–2015"; David Siders et al., "Jerry Brown Signs California Vaccine Bill," *The Sacramento Bee*, June 30, 2015.

69. Pro-MED mail, "Measles Update," Sept. 19, 2011.
70. Philippa Roxby, "Measles Outbreak Warning as Cases Rise in Europe and UK," BBC News, May 13, 2011.
71. Pro-MED mail, "Measles Update."
72. "WHO: Europe Must Act on Measles Outbreak," Dec. 2, 2011, www.telegraph.co.uk.
73. Susana Ferreira, "Cholera Fallout: Can Haitians Sue the U.N. for the epidemic?" *Time*, Dec. 13, 2011.
74. Interview with Mario Joseph, Aug. 14, 2013.
75. R. S. Hendriksen et al., "Population Genetics of *Vibrio cholerae* from Nepal in 2010: Evidence on the Origin of the Haitian Outbreak," *mBio* 2, no. 4 (2011): e00157-11.

7. THE CURE

1. Robert A. Phillips, "The Patho-Physiology of Cholera," *Bulletin of the World Health Organization* 28, no. 3 (1963): 297.
2. Delaporte, *Disease and Civilization*, 88, 90.
3. Chambers, *The Conquest of Cholera*, 168.
4. David Wootton, *Bad Medicine: Doctors Doing Harm Since Hippocrates* (New York: Oxford University Press, 2006).
5. Travis Proulx, Michael Inzlicht, and Eddie Harmon-Jones, "Understanding All Inconsistency Compensation as a Palliative Response to Violated Expectations," *Trends in Cognitive Sciences* 16, no. 5 (2012): 285–91.
6. Thomas S. Kuhn, *The Structure of Scientific Revolutions*, 4th ed. (Chicago: University of Chicago Press, 2012).
7. Wootton, *Bad Medicine*.
8. Ibid.
9. B. A. Foëx, "How the Cholera Epidemic of 1831 Resulted in a New Technique for Fluid Resuscitation," *Emergency Medicine Journal* 20, no. 4 (2003): 316–18.
10. Walter J. Daly and Herbert L. DuPont, "The Controversial and Short-Lived Early Use of Rehydration Therapy for Cholera," *Clinical Infectious Diseases* 47, no. 10 (2008): 1315–19.
11. James Johnson, ed., *The Medico-Chirurgical Review*, vol. 21, 1832.
12. Daly and DuPont, "The Controversial and Short-Lived Early Use of Rehydration Therapy for Cholera."
13. Anthony R. Mawson, "The Hands of John Snow: Clue to His Untimely Death?" *Journal of Epidemiology & Community Health* 63, no. 6 (2009): 497–99.
14. David E. Lilienfeld, "John Snow: The First Hired Gun?" *American Journal of Epidemiology* 152, no. 1 (2000): 4–9; Johnson, *The Ghost Map*, 67.
15. Mawson, "The Hands of John Snow."
16. S.W.B. Newsom, "Pioneers in Infection Control: John Snow, Henry Whitehead, the Broad Street Pump, and the Beginnings of Geographical Epidemiology," *The Journal of Hospital Infection* 64, no. 3 (2006): 210–16.
17. Nigel Paneth et al., "A Rivalry of Foulness: Official and Unofficial Investigations of the London Cholera Epidemic of 1854," *American Journal of Public Health* 88, no. 10 (1998): 1545–53.
18. Lilienfeld, "John Snow."
19. Ibid.
20. Mawson, "The Hands of John Snow."

21. Lilienfeld, "John Snow."

22. Richard L. Guerrant, Benedito A. Carneiro-Filho, and Rebecca A. Dillingham, "Cholera, Diarrhea, and Oral Rehydration Therapy: Triumph and Indictment," *Clinical Infectious Diseases* 37, no. 3 (2003): 398–405.

23. Rosenberg, *The Cholera Years*, 184.

24. Porter, *The Greatest Benefit*, 266.

25. John S. Haller, "Samson of the Materia Medica: Medical Theory and the Use and Abuse of Calomel: In Nineteenth Century America," *Pharmacy in History* 13, no. 2 (1971): 67–76.

26. Wootton, *Bad Medicine*.

27. Thomas W. Clarkson, "The Toxicology of Mercury," *Critical Reviews in Clinical Laboratory Sciences* 34, no. 4 (1997): 369–403.

28. B. S. Drasar and D. Forrest, eds., *Cholera and the Ecology of "Vibrio cholerae"* (London: Chapman & Hall, 1996), 55.

29. Stephen Halliday, *The Great Stink: Sir Joseph Bazalgette and the Cleansing of the Victorian Metropolis* (Mount Pleasant, SC: History Press, 2003); Dale H. Porter, *The Life and Times of Sir Goldsworthy Gurney: Gentleman Scientist and Inventor, 1793–1875* (Bethlehem, PA: Lehigh University Press, 1998).

30. John D. Thompson. "The Great Stench or the Fool's Argument," *The Yale Journal of Biology and Medicine* 64, no. 5 (1991): 529.

31. Halliday, *The Great Stink*; Johnson, *The Ghost Map*, 120; Solomon, *Water*, 258.

32. Kuhn, *The Structure of Scientific Revolutions*.

33. Porter, *Greatest Benefit*, 57.

34. Comment by David Fisman, Feb. 10, 2015.

35. Wootton, *Bad Medicine*.

36. Ibid.

37. Echenberg, *Africa in the Time of Cholera*, 31.

38. Porter, *The Life and Times of Sir Goldsworthy Gurney*.

39. Ibid.

40. Ibid.

41. Thompson, "The Great Stench or the Fool's Argument."

42. Halliday, *The Great Stink*.

43. "Location of Parliaments in the 13th Century," www.parliament.uk.

44. David Boswell Reid, *Ventilation in American Dwellings* (New York: Wiley & Halsted, 1858).

45. Robert Bruegmann, "Central Heating and Forced Ventilation: Origins and Effects on Architectural Design," *Journal of the Society of Architectural Historians* 37, no. 3 (Oct. 1978): 143–60.

46. Thompson, "The Great Stench or the Fool's Argument."

47. Halliday, *The Great Stink*.

48. Porter, *The Life and Times of Sir Goldsworthy Gurney*.

49. Koeppel, *Water for Gotham*, 141.

50. Blake, *Water for the Cities*, 171.

51. Koeppel, *Water for Gotham*, 287.

52. Duffy, *A History of Public Health*, 398, 418.

53. Rosenberg, *The Cholera Years*, 184; Allen, "5 Points Had Good Points."

54. Snowden, *Naples in the Time of Cholera*, 190.

55. Evans, *Death in Hamburg*, 292.

56. Snowden, *Naples in the Time of Cholera*, 69, 100, 190.

57. Evans, *Death in Hamburg*.

58. Nicholas Bakalar, "Milestones in Combating Cholera," *The New York Times*, Oct. 1, 2012.

59. Norman Howard-Jones, "Gelsenkirchen Typhoid Epidemic of 1901, Robert Koch, and the Dead Hand of Max von Pettenkofer," *BMJ* 1, no. 5845 (1973): 103.

60. Alfred S. Evans, "Pettenkofer Revisited: The Life and Contributions of Max von Pettenkofer (1818–1901)," *The Yale Journal of Biology and Medicine* 46, no. 3 (1973): 161; Alfred S. Evans, "Two Errors in Enteric Epidemiology: The Stories of Austin Flint and Max von Pettenkofer," *Review of Infectious Diseases* 7, no. 3 (1985): 434–40.

61. Echenberg, *Africa in the Time of Cholera*, 9.

62. Evans, *Death in Hamburg*, 497–98; Evans, "Two Errors in Enteric Epidemiology"; Christopher Hamlin, *Cholera: The Biography* (New York: Oxford University Press, 2009), 177.

63. Evans, *Death in Hamburg*, 292.

64. Alfredo Morabia, "Epidemiologic Interactions, Complexity, and the Lonesome Death of Max von Pettenkofer," *American Journal of Epidemiology* 166, no. 11 (2007): 1233–38.

65. Melosi, *The Sanitary City*, 94; S. J. Burian et al., "Urban Wastewater Management in the United States: Past, Present, and Future," *Journal of Urban Technology* 7 (2000): 33–62.

66. Ewald, *Evolution of Infectious Disease*, 72–73.

67. Hamlin, *Cholera*, 242.

68. Guerrant, "Cholera, Diarrhea, and Oral Rehydration Therapy."

69. Katherine Harmon, "Can a Vaccine Cure Haiti's Cholera?" *Scientific American*, Jan. 12, 2012.

70. Anwar Huq et al., "Simple Sari Cloth Filtration of Water Is Sustainable and Continues to Protect Villagers from Cholera in Matlab, Bangladesh," *mBio* 1, no. 1 (2010): e00034-10.

71. S. Fannin et al., "A Cluster of Kaposi's Sarcoma and Pneumocystis Carinii Pneumonia Among Homosexual Male Residents of Los Angeles and Range Counties, California," *MMWR* 31, no. 32 (June 18, 1982): 305–307.

72. Charlie Cooper, "Ebola Outbreak: Why Has 'Big Pharma' Failed Deadly Virus' Victims?" *The Independent*, Sept. 7, 2014.

73. Marc H. V. Van Regenmortel, "Reductionism and Complexity in Molecular Biology," *EMBO Reports* 5, no. 11 (2004): 1016.

74. Andrew C. Ahn et al., "The Limits of Reductionism in Medicine: Could Systems Biology Offer an Alternative?" *PLoS Medicine* 3, no. 6 (2006): e208.

75. Laura H. Kahn, "Confronting Zoonoses, Linking Human and Veterinary Medicine," *Emerging Infectious Diseases* 12, no. 4 (2006): 556.

76. Ewan M. Harrison et al., "A Shared Population of Epidemic Methicillin-Resistant *Staphylococcus aureus* 15 Circulates in Humans and Companion Animals," *mBio* 5, no. 3 (2014): e00985-13.

77. Mathieu Albert et al., "Biomedical Scientists' Perception of the Social Sciences in Health Research," *Social Science & Medicine* 66, no. 12 (2008): 2520–31.

78. Interview with Dr. Larry Hribar, Feb. 8, 2012; "More than 1,000 Exposed to Dengue in Florida: CDC," Reuters, July 13, 2010.

8. THE REVENGE OF THE SEA
 1. Sonia Shah, *Crude: The Story of Oil* (New York: Seven Stories Press, 2004), 161.
 2. Environmental Protection Agency, "Climate Change Indicators in the United States: Ocean Heat," Oct. 29, 2014.
 3. Rachel Carson, *The Sea Around Us* (New York: Oxford University Press, 1951), ix.
 4. Sir Alister Hardy, *Great Waters: A Voyage of Natural History to Study Whales, Plankton, and the Waters of the Southern Ocean* (New York: Harper, 1967).
 5. R. R. Colwell, J. Kaper, and S. W. Joseph, "*Vibrio cholerae, Vibrio parahaemolyticus*, and Other *Vibrios*: Occurrence and Distribution in Chesapeake Bay," *Science*, 198, no. 4315 (Oct. 28, 1977): 394–96.
 6. Interview with Rita Colwell.
 7. Anwar Huq, R. Bradley Sack, and Rita Colwell, "Cholera and Global Ecosystems," in Aron and Patz, *Ecosystem Change and Public Health*, 333.
 8. Arnold Taylor, "Plankton and the Gulf Stream," *New Scientist*, March 1991.
 9. Huq, Sack, and Colwell, "Cholera and Global Ecosystems," 336; Luigi Vezzulli, Rita R. Colwell, and Carla Pruzzo, "Ocean Warming and Spread of Pathogenic Vibrios in the Aquatic Environment," *Microbial Ecology* 65, no. 4 (2013): 817–25; Graeme C. Hays, Anthony J. Richardson, and Carol Robinson, "Climate Change and Marine Plankton," *Trends in Ecology & Evolution* 20, no. 6 (2005): 337–44; Gregory Beaugrand, Luczak Christophe, and Edwards Martin, "Rapid Biogeographical Plankton Shifts in the North Atlantic Ocean," *Global Change Biology* 15, no. 7 (2009): 1790–1803.
10. William H. McNeill, *Plagues and Peoples* (Garden City, NY: Anchor Press, 1976), 283.
11. Oscar Felsenfeld, "Some Observations on the Cholera (El Tor) Epidemic in 1961–62," *Bulletin of the World Health Organization* 28, no. 3 (1963): 289–96.
12. Ibid.
13. Rudolph Hugh, "A Comparison of *Vibrio cholerae* Pacini and *Vibrio eltor* Pribram," *International Bulletin of Bacteriological Nomenclature and Taxonomy* 15, no. 1 (1965): 61–68.
14. Paul H. Kratoska, ed., *Southeast Asia Colonial History: High Imperialism (1890s–1930s)* (New York: Routledge, 2001).
15. C. E. de Moor, "Paracholera (El Tor): Enteritis Choleriformis El Tor van Loghem," *Bulletin of the World Health Organization* 2 (1949): 5–17.
16. Agus P. Sari et al., "Executive Summary: Indonesia and Climate Change: Working Paper on Current Status and Policies," Department for International Development and the World Bank, March 2007; Bernhard Glaeser and Marion Glaser, "Global Change and Coastal Threats: The Indonesian Case. An Attempt in Multi-Level Social-Ecological Research," *Human Ecology Review* 17, no. 2 (2010); Kathleen Schwerdtner Máñez et al., "Water Scarcity in the Spermonde Archipelago, Sulawesi, Indonesia: Past, Present and Future," *Environmental Science & Policy* 23 (2012): 74–84.
17. Felsenfeld, "Some Observations on the Cholera (El Tor) Epidemic."

18. "Far East Pressing Anti-Cholera Steps," *The New York Times*, Aug. 20, 1961; "Chinese Reds Blame U.S. in Cholera Rise," *The New York Times*, Aug. 19, 1961.

19. C. Sharma et al., "Molecular Evidence That a Distinct *Vibrio cholerae* 01 Biotype El Tor Strain in Calcutta May Have Spread to the African Continent," *Journal of Clinical Microbiology* 36, no. 3 (March 1998): 843–44.

20. Echenberg, *Africa in the Time of Cholera*, 125–27.

21. Oscar Felsenfeld, "Present Status of the El Tor Vibrio Problem," *Bacteriological Reviews* 28, no. 1 (1964): 72; Colwell, "Global Climate and Infectious Disease."

22. Iván J. Ramírez, Sue C. Grady, and Michael H. Glantz, "Reexamining El Niño and Cholera in Peru: A Climate Affairs Approach," *Weather, Climate, and Society* 5 (2013): 148–61.

23. Bill Manson, "The Ocean Has a Long Memory," *San Diego Reader*, Feb. 12, 1998; Rosa R. Mouriño-Pérez, "Oceanography and the Seventh Cholera Pandemic," *Epidemiology* 9, no. 3 (1998): 355–57.

24. Ramírez, Grady, and Glantz, "Reexamining El Niño and Cholera in Peru"; María Ana Fernández-Álamo and Jaime Färber-Lorda, "Zooplankton and the Oceanography of the Eastern Tropical Pacific: A Review," *Progress in Oceanography* 69, no. 2 (2006): 318–59; Bert Rein et al., "El Niño Variability off Peru During the Last 20,000 Years," *Paleoceanography* 20, no. 4 (2005); Jaime Martinez-Urtaza et al., "Emergence of Asiatic Vibrio Diseases in South America in Phase with El Niño," *Epidemiology* 19, no. 6 (2008): 829–37.

25. Vezzulli, Colwell, and Pruzzo, "Ocean Warming and Spread of Pathogenic Vibrios"; Rafael Montilla et al., "Serogroup Conversion of Vibrio Cholerae non-O1 to Vibrio Cholerae O1: Effect of Growth State of Cells, Temperature, and Salinity," *Canadian Journal of Microbiology* 42, no. 1 (1996): 87–93; Luigi Vezzulli et al., "Dual Role Colonization Factors Connecting *Vibrio cholerae*'s Lifestyles in Human and Aquatic Environments Open New Perspectives for Combating Infectious Diseases," *Current Opinions in Biotechnology* 19 (2008): 254–59.

26. P. R. Epstein, "Algal Blooms in the Spread and Persistence of Cholera," *BioSystems* 31, no. 2 (1993): 209–221; Jeffrey W. Turner et al., "Plankton Composition and Environmental Factors Contribute to Vibrio Seasonality," *The ISME Journal* 3, no. 9 (2009): 1082–92.

27. Connie Lam et al., "Evolution of Seventh Cholera Pandemic and Origin of 1991 Epidemic, Latin America," *Emerging Infectious Diseases* 16, no. 7 (2010): 1130.

28. "Cholera Epidemic Kills 51 in Peru," *The Times* (London), Feb. 11, 1991.

29. Simon Strong, "Peru Minister Quits in Cholera Row," *The Independent*, March 19, 1991; Malcolm Coad, "Peru's Cholera Epidemic Spreads to Its Neighbors," *The Guardian*, April 18, 1991; "Cholera Cases Confirmed Near Border with U.S.," *Montreal Gazette*, March 18, 1992; William Booth, "Cholera's Mysterious Journey North," *The Washington Post*, Aug. 26, 1991; "Baywatch Filming Hit by Cholera Alert," *London Evening Standard*, July 29, 1992; Barbara Turnbull, "Flight Hit by Cholera, 2 Sought in Canada," *Toronto Star*, Feb. 22, 1992; Les Whittington, "Mexico; Traffickers Blamed for Spread of Cholera," *Ottawa Citizen*, Sept. 11, 1991.

30. J. P. Guthmann, "Epidemic Cholera in Latin America: Spread and Routes of Transmission," *The Journal of Tropical Medicine and Hygiene* 98, no. 6 (1995): 419.

31. Jazel Dolores and Karla J. F. Satchell, "Analysis of *Vibrio cholerae*: Genome Sequences Reveals Unique rtxA Variants in Environmental Strains and an rtxA-Null

Mutation in Recent Altered El Tor Isolates," *mBio* 4, no. 2 (2013); Ashrafus Safa, G. Balakrish Nair, and Richard Y. C. Kong, "Evolution of New Variants of *Vibrio cholerae* O1," *Trends in Microbiology* 18, no. 1 (2010): 46–54.

32. A. K. Siddique et al., "El Tor Cholera with Severe Disease: A New Threat to Asia and Beyond," *Epidemiology and Infection* 138, no. 3 (2010): 347–52.

33. R. Piarroux and B. Faucher, "Cholera Epidemics in 2010: Respective Roles of Environment, Strain Changes, and Human-Driven Dissemination," *Clinical Microbiology and Infection* 18, no. 3 (2012): 231–38.

34. Deborah Jenson et al., "Cholera in Haiti and Other Caribbean Regions, 19th Century," *Emerging Infectious Diseases* 17, no. 11 (Nov. 2011).

35. Interview with Anwar Huq, Jan. 25, 2011.

36. Interview with Rita Colwell; "The United Nations' Duty in Haiti's Cholera Outbreak," *The Washington Post*, Aug. 11, 2013.

37. Carlos Seas et al., "New Insights on the Emergence of Cholera in Latin America During 1991: the Peruvian Experience," *American Journal of Tropical Medicine and Hygiene* 62, no. 4 (2000): 513–17.

38. Luigi Vezzulli et al., "Long-Term Effects of Ocean Warming on the Prokaryotic Community: Evidence from the Vibrios," *The ISME Journal* 6, no. 1 (2012): 21–30.

39. Peter Andrey Smith, "Sea Sick," *Modern Farmer*, Sept. 11, 2013.

40. Colwell, "Global Climate and Infectious Disease."

41. Alexander, "An Overview of the Epidemiology of Avian Influenza."

42. Drexler, *Secret Agents*, 65.

43. Joan Brunkard, "Climate Change Impacts on Waterborne Diseases Outbreaks," International Conference on Emerging Infectious Diseases, Atlanta, GA, March 12, 2012; Violeta Trinidad Pardío Sedas, "Influence of Environmental Factors on the Presence of *Vibrio cholerae* in the Marine Environment: A Climate Link," *The Journal of Infection in Developing Countries* 1, no. 3 (2007): 224–41.

44. Jonathan E. Soverow et al., "Infectious Disease in a Warming World: How Weather Influenced West Nile Virus in the United States (2001–2005)," *Environmental Health Perspectives* 117, no. 7 (2009): 1049–52.

45. Peter Daszak, "Fostering Advances in Interdisciplinary Climate Science," lecture, Arthur M. Sackler Colloquia of the National Academy of Sciences, Washington, DC, March 31–April 2, 2011.

46. S. Mistry and A. Moreno-Valdez, "Climate Change and Bats: Vampire Bats Offer Clues to the Future," *Bats* 26, no. 2 (Summer 2008).

47. Lars Eisen and Chester G. Moore, "*Aedes* (*Stegomyia*) *aegypti* in the Continental United States: a Vector at the Cool Margin of Its Geographic Range," *Journal of Medical Entomology* 50, no. 3 (2013): 467–78; Diana Marcum, "California Residents Cautioned to Look Out for Yellow Fever Mosquito," *Los Angeles Times*, Oct. 20, 2013.

48. D. Roiz et al., "Climatic Factors Driving Invasion of the Tiger Mosquito (*Aedes albopictus*) into New Areas of Trentino, Northern Italy," *PLoS ONE* 6, no. 4 (April 15, 2011): e14800.

49. Laura Jensen, "What Does Climate Change and Deforestation Mean for Lyme Disease in the 21st Century?" Tick Talk, an investigative project on Lyme disease, SUNY New Paltz.

50. Andrew Nikiforuk, "Beetlemania," *New Scientist*, Nov. 5, 2011.
51. M. C. Fisher et al., "Emerging Fungal Threats to Animal, Plant and Ecosystem Health," *Nature* 484 (April 2012): 186–94.
52. Ibid.
53. Arturo Casadevall, "Fungi and the Rise of Mammals," *PLoS Pathogens* 8, no. 8 (2012): e1002808.
54. Arturo Casadevall, "Thoughts on the Origin of Microbial Virulence," International Conference on Emerging Infectious Diseases, Atlanta, GA, March 13, 2012.
55. Letter from Larry Madoff to Pro-MED mail subscribers, June 5, 2012.
56. Fisher, "Emerging Fungal Threats to Animal, Plant and Ecosystem Health."

9. THE LOGIC OF PANDEMICS

1. Markus G. Weinbauer and Fereidoun Rassoulzadegan, "Extinction of Microbes: Evidence and Potential Consequences," *Endangered Species Research* 3, no. 2 (2007): 205–15; Gerard Tortora, Berdelle Funke, and Christine Case, *Microbiology: An Introduction*, 10th ed. (San Francisco: Pearson Education, 2010).
2. Kat McGowan, "How Life Made the Leap from Single Cells to Multicellular Animals," *Wired*, Aug. 1, 2014.
3. Blood samples taken from subjects who viewed pictures of people sneezing or with pox lesions had 23.6 percent more interleukin-6 than samples taken from subjects who viewed pictures of pointed guns and furniture. C. L. Fincher and R. Thornhill, "Parasite-Stress Promotes In-Group Assortative Sociality: The Cases of Strong Family Ties and Heightened Religiosity," *Behavioral and Brain Sciences* 35, no. 2 (2012): 61–79.
4. Sabra L. Klein and Randy J. Nelson, "Influence of Social Factors on Immune Function and Reproduction," *Reviews of Reproduction* 4, no. 3 (1999): 168–78.
5. Matt Ridley, *The Red Queen: Sex and the Evolution of Human Nature* (New York: Macmillan, 1994), 80.
6. Michael A. Brockhurst, "Sex, Death, and the Red Queen," *Science*, July 8, 2011.
7. Makoto Takeo et al., "Wnt Activation in Nail Epithelium Couples Nail Growth to Digit Regeneration," *Nature* 499, no. 7457 (2013): 228–32.
8. Joshua Mitteldorf, "Evolutionary Origins of Aging," in Gregory M. Fahy et al., eds., *The Future of Aging: Pathways to Human Life Extension* (Dordrecht: Springer, 2010).
9. Jerome Wodinsky, "Hormonal Inhibition of Feeding and Death in Octopus: Control by Optic Gland Secretion," *Science* 198, no. 4320 (1977): 948–51.
10. Valter D. Longo, Joshua Mitteldorf, and Vladimir P. Skulachev, "Programmed and Altruistic Ageing," *Nature Reviews Genetics* 6, no. 11 (2005): 866–72.
11. Interview with Joshua Mitteldorf, Feb. 4, 2015.
12. Catherine Clabby, "A Magic Number? An Australian Team Says It Has Figured Out the Minimum Viable Population for Mammals, Reptiles, Birds, Plants and the Rest," *American Scientist* 98 (2010): 24–25.
13. Curtis H. Flather et al., "Minimum Viable Populations: Is There a 'Magic Number' for Conservation Practitioners?" *Trends in Ecology & Evolution* 26, no. 6 (2011): 307–16.
14. According to the adaptive theory of aging, suicide genes are adaptive for groups rather than individuals. The precise evolutionary mechanisms by which so-called

group selection could occur are unclear. Joshua Mitteldorf and John Pepper, "Senescence as an Adaptation to Limit the Spread of Disease," *Journal of Theoretical Biology* 260, no. 2 (2009): 186–95.

15. Diogo Meyer and Glenys Thomson, "How Selection Shapes Variation of the Human Major Histocompatibility Complex: A Review," *Annals of Human Genetics* 65, no. 1 (2001): 1–26.

16. Interview with Glenys Thomson, Feb. 6, 2015; Meyer and Thomson, "How Selection Shapes Variation of the Human Major Histocompatibility Complex."

17. Ajit Varki, "Human-Specific Changes in Siglec Genes," video lecture, CARTA: The Genetics of Humanness, April 9, 2011; Darius Ghaderi et al., "Sexual Selection by Female Immunity Against Paternal Antigens Can Fix Loss of Function Alleles," *Proceedings of the National Academy of Sciences* 108, no. 43 (2011): 17743–48.

18. Alasdair Wilkins, "How Sugar Molecules Secretly Shaped Human Evolution," io9, Oct. 10, 2011.

19. Interview with Ajit Varki, Feb. 9, 2015; Bruce Lieberman, "Human Evolution: Details of Being Human," *Nature*, July 2, 2008.

20. Kenneth D. Beaman et al., "Immune Etiology of Recurrent Pregnancy Loss and Its Diagnosis," *American Journal of Reproductive Immunology* 67, no. 4 (2012): 319–25.

21. Annie N. Samraj et al., "A Red Meat–Derived Glycan Promotes Inflammation and Cancer Progression," *Proceedings of the National Academy of Sciences* 112, no. 2 (2015): 542–47.

22. F. B. Piel et al., "Global Epidemiology of Sickle Haemoglobin in Neonates: A Contemporary Geostatistical Model-Based Map and Population Estimates," *The Lancet* 381, no. 9861 (Jan. 2013): 142–51.

23. Elinor K. Karlsson, Dominic P. Kwiatkowski, and Pardis C. Sabeti, "Natural Selection and Infectious Disease in Human Populations," *Nature Reviews Genetics* 15, no. 6 (2014): 379–93.

24. David J. Anstee, "The Relationship Between Blood Groups and Disease," *Blood* 115, no. 23 (2010): 4635–43.

25. Karlsson, Kwiatkowski, and Sabeti, "Natural Selection and Infectious Disease in Human Populations."

26. Anstee, "The Relationship Between Blood Groups and Disease."

27. Gregory Demas and Randy Nelson, eds., *Ecoimmunology* (New York: Oxford University Press, 2012), 234.

28. Meyer and Thomson, "How Selection Shapes Variation of the Human Major Histocompatibility Complex."

29. Fincher and Thornhill, "Parasite-Stress Promotes In-Group Assortative Sociality."

30. McNeill, *Plagues and Peoples*, 91–92.

31. Fincher and Thornhill, "Parasite-Stress Promotes In-Group Assortative Sociality."

32. E. Cashdan, "Ethnic Diversity and Its Environmental Determinants: Effects of Climate, Pathogens, and Habitat Diversity," *American Anthropologist* 103 (2001): 968–91.

33. Carlos David Navarrete and Daniel M. T. Fessler, "Disease Avoidance and Ethnocentrism: The Effects of Disease Vulnerability and Disgust Sensitivity on Intergroup Attitudes," *Evolution and Human Behavior* 27, no. 4 (2006): 270–82.

34. Andrew Spielman and Michael d'Antonio, *Mosquito: The Story of Man's Deadliest Foe* (New York: Hyperion, 2002), 49.

35. Diamond, *Guns, Germs, and Steel*, 210–11.

36. Sonia Shah, *The Fever: How Malaria Has Ruled Humankind for 500,000 Years* (New York: Farrar, Straus and Giroux, 2010), 41–43.

37. R. Thornhill and S. W. Gangestad, "Facial Sexual Dimorphism, Developmental Stability and Susceptibility to Disease in Men and Women," *Evolution and Human Behavior* 27 (2006): 131–44.

38. A. Booth and J. Dabbs, "Testosterone and Men's Marriages," *Social Forces* 72 (1993): 463–77.

39. Anthony C. Little, Lisa M. DeBruine, and Benedict C. Jones, "Exposure to Visual Cues of Pathogen Contagion Changes Preferences for Masculinity and Symmetry in Opposite-Sex Faces," *Proceedings of the Royal Society B: Biological Sciences* 278, no. 1714 (2011): 2032–39.

40. Meyer and Thomson, "How Selection Shapes Variation of the Human Major Histocompatibility Complex."

41. Margaret McFall-Ngai et al., "Animals in a Bacterial World, a New Imperative for the Life Sciences," *Proceedings of the National Academy of Sciences* 110, no. 9 (2013): 3229–36; Gerard Eberl, "A New Vision of Immunity: Homeostasis of the Superorganism," *Mucosal Immunology* 3, no. 5 (2010): 450–60.

42. John F. Cryan and Timothy G. Dinan, "Mind-Altering Microorganisms: The Impact of the Gut Microbiota on Brain and Behaviour," *Nature Reviews Neuroscience* 13, no. 10 (2012): 701–12.

43. McGowan, "How Life Made the Leap from Single Cells to Multicellular Animals."

44. F. Prugnolle et al., "Pathogen-Driven Selection and Worldwide HLA Class I Diversity," *Current Biology* 15 (2005): 1022–27.

45. Kenneth Miller, "Archaeologists Find Earliest Evidence of Humans Cooking with Fire," *Discover*, May 2013.

46. Christopher Sandom et al., "Global Late Quaternary Megafauna Extinctions Linked to Humans, Not Climate Change," *Proceedings of the Royal Society B: Biological Sciences* 281, no. 1787 (June 4, 2014).

10. TRACKING THE NEXT CONTAGION

1. Saeed Ahmed and Dorrine Mendoza, "Ebola Hysteria: An Epic, Epidemic Overreaction," CNN, Oct. 20, 2014.

2. Reuters, "Kentucky Teacher Resigns Amid Parents' Ebola Fears: Report," *The Huffington Post*, Nov. 3, 2014; Olga Khazan, "The Psychology of Irrational Fear," *The Atlantic*, Oct. 31, 2014; Amanda Terkel, "Oklahoma Teacher Will Have to Quarantine Herself After Trip to Ebola-free Rwanda," *The Huffington Post*, Oct. 28, 2014; Amanda Cuda and John Burgeson, "Milford Girl in Ebola Scare Wants to Return to School," www.CTPost.com, Oct. 30, 2014.

3. Matt Byrne, "Maine School Board Puts Teacher on Leave After She Traveled to Dallas," *Portland Press Herald*, Oct. 17, 2014.

4. Ahmed and Mendoza, "Ebola Hysteria"; CDC, "It's Turkey Time: Safely Prepare Your Holiday Meal," Nov. 25, 2014.

5. Khazan, "The Psychology of Irrational Fear."

6. Jere Longman, "Africa Cup Disrupted by Ebola Concerns," *The New York Times*, Nov. 11, 2014; "The Ignorance Epidemic," *The Economist*, Nov. 15, 2014.

7. Eyder Peralta, "Health Care Worker on Cruise Ship Tests Negative for Ebola," NPR, Oct. 19, 2014.

8. "'Ebola' Coffee Cup Puts Plane on Lockdown at Dublin Airport," RT.com, Oct. 30, 2014.

9. "Ottawa's Ebola Overkill," *The Globe and Mail*, Nov. 3, 2014.

10. Drew Hinshaw and Jacob Bunge, "U.S. Buys Up Ebola Gear, Leaving Little for Africa," *The Wall Street Journal*, Nov. 25, 2014.

11. Katie Helper, "More Americans Have Been Married to Kim Kardashian than Have Died from Ebola," *Raw Story*, Oct. 22, 2014.

12. H. Rhee and D. J. Cameron, "Lyme Disease and Pediatric Autoimmune Neuropsychiatric Disorders Associated with Streptococcal Infections (PANDAS): An Overview," *International Journal of General Medicine* 5 (2012): 163–74.

13. Jennifer Newman, "Local Lyme Impacts Outdoor Groups and Businesses," and Zameena Mejia, "On the Trail of De-Railing Lyme," Tick Talk, State University of New York at New Paltz, 2014.

14. Maria G. Guzman, Mayling Alvarez, and Scott B. Halstead, "Secondary Infection as a Risk Factor for Dengue Hemorrhagic Fever/Dengue Shock Syndrome: An Historical Perspective and Role of Antibody-Dependent Enhancement of Infection," *Archives of Virology* 158, no. 7 (2013): 1445–59; "Dengue," CDC website, June 9, 2014.

15. Sean Kinney, "CDC Errs in Levels of Dengue Cases in Key West," *Florida Keys Keynoter*, July 17, 2010.

16. Sean Kinney, "CDC Stands by Key West Dengue-Fever Report," *Florida Keys Keynoter*, July 28, 2010.

17. Denise Grady and Catharine Skipp, "Dengue Fever? What About It, Key West Says," *The New York Times*, July 24, 2010.

18. Bob LaMendola, "Broward Woman Gets Dengue Fever on Key West Trip," *Sun-Sentinel*, July 30, 2010.

19. "Woman in Florida Diagnosed with Cholera," CNN, Nov. 17, 2010; "Cholera, Diarrhea and Dysentery Update 2011 (23): Haiti, Dominican Republic," ProMED, July 26, 2011; Juan Tamayo, "Cholera Reportedly Kills 15, Sickens Hundreds in Eastern Cuba," *The Miami Herald*, July 6, 2012; Fox News Latino, "Puerto Rico: Cholera, After Affecting Haiti and Dominican Republic, Hits Island," July 5, 2011; "Shanty Towns and Cholera," editorial, *The Freeport News*, Nov. 15, 2012.

20. "Why Pandemic Disease and War are So Similar," *The Economist*, March 28, 2015.

21. Deborah A. Adams et al., "Summary of Notifiable Diseases—United States, 2011," *MMWR* 60, no. 53 (July 5, 2013): 1–117.

22. Stephen S. Morse, "Public Health Surveillance and Infectious Disease Detection," *Biosecurity and Bioterrorism* 10, no. 1 (2012): 6–16.

23. Baize, "Emergence of Zaire Ebola Virus Disease in Guinea."

24. Norimitsu Onishi, "Empty Ebola Clinics in Liberia Are Seen as Misstep in US Relief Effort," *The New York Times*, April 11, 2015.

25. Interview with Leo Poon, Hong Kong, Jan. 2012.

26. Karen J. Monaghan, "SARS: Down but Still a Threat," in Institute of Medicine, *Learning from SARS: Preparing for the Next Disease Outbreak* (Washington, DC: National Academies Press, 2004), 255.

27. Erin Place, "In Light of EEE Death, County Opts to Spray," *The Palladium-Times*, Aug. 16, 2011.
28. Interview with Ivan Gayton, June 26, 2014.
29. Aleszu Bajak, "Asian Tiger Mosquito Could Expand Painful Caribbean Virus into U.S.," *Scientific American*, Aug. 12, 2014; Pan American Health Organization, "Chikungunya: A New Virus in the Region of the Americas," July 8, 2014.
30. Charles Kenny, "The Ebola Outbreak Shows Why the Global Health System Is Broken," *BusinessWeek*, Aug. 11, 2014; Kus, "New Delhi Metallo-ss-lactamase-1"; Interview with Malik Peiris; Davis, *The Monster at Our Door*, 112.
31. Interview with Leo Poon.
32. USAID, "Emerging Pandemic Threats: Program Overview," June 2010.
33. Martin Cetron, "Clinician-Based Surveillance Networks Utilizing Travelers as Sentinels for Emerging Infectious Diseases," International Conference on Emerging Infectious Diseases, Atlanta, GA, March 13, 2012.
34. Interview with James Wilson, July 31, 2013; Wolfe, *The Viral Storm*, 213; Rodrique Ngowi, "US Bots Flagged Ebola Before Outbreak Announced," Associated Press, Aug. 9, 2014.
35. Interview with James Wilson; Wolfe, *The Viral Storm*, 195, 213; Ngowi, "US Bots Flagged Ebola Before Outbreak Announced"; David Braun, "Anatomy of the Discovery of the Deadly Bas-Congo Virus," *National Geographic*, Sept. 27, 2012.
36. Gina Kolata, "The New Generation of Microbe Hunters," *The New York Times*, Aug. 29, 2011; Jan Semenza, "The Impact of Economic Crises on Communicable Diseases," International Conference on Emerging Infectious Diseases, Atlanta, GA, March 12, 2012.
37. Larry Brilliant, "My Wish: Help Me Stop Pandemics," TED, Feb. 2006.
38. Interview with Peter Daszak.
39. Walsh, "Emerging Carbapenemases."
40. Alex Whiting, "New Pandemic Insurance to Prevent Crises Through Early Payouts," Reuters, March 26, 2015.
41. Interview with James Wilson.
42. Christopher Joyce, "Cellphones Could Help Doctors Stay Ahead of an Epidemic," *Shots*, NPR's Health Blog, Aug. 31, 2011.
43. Pan American Health Organization, "Epidemiological Update: Cholera," March 20, 2014.
44. Belle-Anse is not unique in suffering the effects of poorly maintained aid projects. The entire country is littered with them. According to a 2012 survey, more than a third of the wells in Haiti constructed by aid groups—most of which are left unmanaged—are contaminated with fecal bacteria. After returning to Port-au-Prince, I met a young British man who told me with pride that he was using his trust fund money to install toilets at a local school. But despite the fact of the ongoing cholera epidemic and the manifest reality that Haitians were regularly exposed to fecal contamination in the environment, he hadn't considered the issue of where the toilets would deposit their contents. When I asked him, he paused. "In the river, I guess," he finally said. "Like everyone else!" See Jocelyn M. Widmer et al., "Water-Related Infrastructure in a Region of Post-Earthquake Haiti: High Levels of Fecal Contamination and Need for Ongoing Monitoring," *The American Journal of Tropical Medicine and Hygiene* 91, no. 4 (Oct. 2014): 790–97.

ACKNOWLEDGMENTS

My sources for *Pandemic* varied widely, from sanitation engineers and archaeologists to geneticists and epidemiologists, but they all had one thing in common, which is a willingness to talk to a science journalist who called up out of the blue. While I have cited only the handful of sources whose words I quote directly, they were all instrumental in the making of this book. Without their generosity, it could never have been written.

The research and reporting I conducted over the past six years would have been impossible, too, without the support of a number of individuals and organizations. The Pulitzer Center on Crisis Reporting, besides supporting my reporting trips, helped turn my narratives of the cholera epidemics in 1832 New York City and 2010 Port-au-Prince into a dazzling interactive visualization ("Mapping Cholera," at choleramap .pulitzercenter.org). This parallel project on the two epidemics punctuating cholera's pandemic sweep crystallized my understanding of the social and political roots of pandemics. Peter Sawyer, Dan McCarey, Nathalie Applewhite, Zach Child, Jon Sawyer, and the rest of the team at the center made it possible. Oliver Schulz and Ivan Gayton at Médecins Sans Frontières, in addition to being heroes on the front lines fighting epidemics, tackled technical and bureaucratic hurdles on multiple continents to allow me to use the remarkable data they collected on cholera in Haiti. Randi Hutter Epstein, Matthew Knutzen, Steven

Romalewski, Don Boyes, and others provided critical early help as well, and the New York Academy of Medicine hosted a public event in connection with the project and this book.

Jim and Mary Ottaway and Lisa Phillips at SUNY–New Paltz offered me an honorary journalism professorship, which allowed me to teach a class on investigating epidemics. The semester-long collaborative investigative project on Lyme disease that resulted was instrumental in my understanding of that cryptic spillover disease. I thank them, and all my students, whose hard work made it possible. Nassim Assefi and her colleagues at TEDMED offered me their well-lit stage to showcase my ideas about pandemics and how we understand them. Jodi Solomon and her staff provided me with the opportunity to present the material in this book to thoughtful audiences across the country.

I owe a huge debt to the journalist Philippe Rivière, formerly of *Le Monde Diplomatique*, who not only designed the lovely maps featured here but also provided critical feedback. Thoughtful comments from my dear friend Michelle Markley and my parents, Dr. Hasmukh Shah and Dr. Hansa Shah, also immensely improved this book. Thanks, too, to David Fisman, who generously made time to review an early draft, and to Michael Olesen, Dao Tran, and Trent Duffy, who offered helpful comments. Frances Botkin, who accompanied me to Haiti, made a difficult trip much less so; Jennifer Ballengee listened attentively to my long harangues as I wrote draft after draft. *Scientific American* magazine, *Yale Environment 360*, *Foreign Affairs*, *The Atlantic*, and *Le Monde Diplomatique* were among the outlets that published articles that supported the reporting on which this book is based. David Fisman and Ashleigh Tuite put together the story of cholera's spread along the Erie Canal and shared their data with me. Su Dongxia provided cheerful logistical support in Guangzhou, as did Rita Choksi in New Delhi and Sean Roubens Jean Sacra in Port-au-Prince. Catherine Guenther provided research assistance.

I thank my lovely agent, Charlotte Sheedy, who gifted me with her unwavering support, and my editor, Sarah Crichton, and the rest of the team at Farrar, Straus and Giroux, who brought this book into the world. Finally, for sustaining me through the years of reporting and writing, I thank Mark Bulmer and our sons, Z and K.

INDEX